SEED ACTIVISM

Food, Health, and the Environment

Series Editors: Robert Gottlieb, Henry R. Luce Professor of Urban and Environmental Policy, Occidental College; Nevin Cohen, Associate Professor, City University of New York (CUNY) Graduate School of Public Health

A complete list of books published in the Food, Health, and the Environment series appears in the back of this book.

SEED ACTIVISM

PATENT POLITICS AND LITIGATION IN THE GLOBAL SOUTH

KARINE E. PESCHARD

THE MIT PRESS   CAMBRIDGE, MASSACHUSETTS   LONDON, ENGLAND

The MIT Press would like to thank the anonymous peer reviewers who provided comments on drafts of this book. The generous work of academic experts is essential for establishing the authority and quality of our publications. We acknowledge with gratitude the contributions of these otherwise uncredited readers.

This book was set in Stone Serif and Avenir by Westchester Publishing Services. Printed and bound in the United States of America.

Published with the support of the Swiss National Science Foundation. The prepress of this publication was funded by the Swiss National Science Foundation.

Library of Congress Cataloging-in-Publication Data is available.

ISBN: 9780262544641 (paperback); 9780262372220 (pdf); 9780262372237 (epub)

10   9   8   7   6   5   4   3   2   1

To my parents

And to all those who keep sowing seeds,
year after year and despite the difficulties.

The relentless march of intellectual property rights needs to be stopped and questioned.
—United Nations Human Development Report (1999)

CONTENTS

LIST OF ACRONYMS AND ABBREVIATIONS

ABRASEM	Associação Brasileira de Sementes e Mudas (Brazilian Seed Producers Association)
ABSP	Agricultural Biotechnology Support Project
ADI	Ação Direta de Inconstitucionalidade (Direct Action of Unconstitutionality)
APASSUL	Associação dos Produtores e Comerciantes de Sementes e Mudas do Rio Grande do Sul (Seed Association of Rio Grande do Sul)
APROSOJA	Associação de Produtores de Soja (Soybean Producers Association)
APROSOJA-MT	Associação de Produtores de Soja e Milho do Estado de Mato Grosso (Soybean Producers Association of Mato Grosso)
APROSOJA-RS	Associação dos Produtores de Soja do Estado do Rio Grande do Sul (Soybean Producers Association of Rio Grande do Sul)
BJP	Bharatiya Janata Party (Indian People's Party)
BKS	Bharatiya Kisan Sangh (Indian Farmers' Union)
BRASPOV	Associação Brasileira de Obtentores Vegetais (Brazilian Plant Breeders' Association)
Bt	*Bacillus thuringensis*

CBD	Convention on Biological Diversity
CCI	Competition Commission of India
CNA	Confederação Nacional da Agricultura e Pecuária do Brasil (National Agricultural Confederation)
CPT	Comissão Pastoral da Terra (Pastoral Land Commission)
CSP	Cotton Seeds Price (Control) Order
CTNBio	Comissão Técnica Nacional em Biossegurança (National Technical Commission on Biosafety)
EMBRAPA	Empresa Brasileira de Pesquisa Agropecuária (Brazilian Agricultural Research Corporation)
ESG	Environment Support Group
FAMATO	Federação da Agricultura e Pecuária do Estado de Mato Grosso (Federation of Agriculture and Livestock of Mato Grosso)
FAO	Food and Agriculture Organization
FARSUL	Federação da Agricultura do Estado do Rio Grande do Sul (Agricultural Federation of the State of Rio Grande do Sul)
FETAG-RS	Federação dos Trabalhadores na Agricultura do Rio Grande do Sul (Federation of Agricultural Workers of Rio Grande do Sul)
GATT	General Agreement on Tariffs and Trade
GEAC	Genetic Engineering Approval Committee (renamed Genetic Engineering Appraisal Committee in 2010)
GMOs	Genetically modified organisms
GoI	Government of India
ICAR	Indian Council of Agricultural Research
INPI	Instituto Nacional da Propriedade Industrial (National Institute of Industrial Property)
INR	Indian rupee
ISAAA	International Service for the Acquisition of Agri-Biotech Applications
ITPGRFA	International Treaty on Plant Genetic Resources for Food and Agriculture ("Plant Treaty")

KBB	Karnataka Biodiversity Board
KIA	Knowledge Initiative on Agriculture
MoEF	Ministry of Environment and Forests
MMB	Mahyco Monsanto Biotech Limited
NBA	National Biodiversity Authority
NSAI	National Seed Association of India
PPVFR Act	Protection of Plant Varieties and Farmers' Rights Act
PVP	Plant variety protection
RR	Roundup Ready
RSS	Rashtriya Swayamsevak Sangh (National Volunteer Organization)
STF	Supremo Tribunal Federal (Federal Supreme Court)
STJ	Supremo Tribunal de Justiça (Superior Court of Justice)
TNAU	Tamil Nadu Agricultural University
TRIPS	Trade-Related Aspects of Intellectual Property Rights
UAS	University of Agricultural Sciences
UPOV	Union Internationale pour la Protection des Obtentions Végétales (International Union for the Protection of New Varieties of Plants)
USAID	US Agency for International Development
WTO	World Trade Organization

LIST OF FIGURES

SERIES FOREWORD

Seed Activism is the twentieth book in the Food, Health, and the Environment series. The series explores the global and local dimensions of food systems and the issues of access, social, environmental and food justice, and community well-being. Books in the series focus on how and where food is grown, manufactured, distributed, sold, and consumed. They address questions of power and control, social movements and organizing strategies, and the health, environmental, social and economic factors embedded in food-system choices and outcomes. As this book demonstrates, the focus is not only on food security and well-being but also on economic, political, and cultural factors together with regional, state, national, and international policy decisions. Food, Health, and the Environment books therefore provide a window into the public debates, alternative and existing discourses, and multidisciplinary perspectives that have made food systems and their connections to both health and the environment critically important subjects of study and for social and policy change.

Robert Gottlieb, Occidental College
Nevin Cohen, City University of New York (CUNY)
Graduate School of Public Health

ACKNOWLEDGMENTS

This book is the outcome of a project grant funded by the Swiss National Science Foundation (SNSF) and coordinated by Shalini Randeria, then director of the Albert Hirschman Centre on Democracy, Graduate Institute of International and Development Studies (IHEID), Geneva, Switzerland. This project would not have been possible without Shalini's support and mentorship.

I also wish to express my gratitude to Christine Lutringer, executive director of the Albert Hirschman Center on Democracy, with whom it has been a real pleasure to work. I thank my colleagues at the Graduate Institute for their comradeship and conversations, in particular Christophe Golay and Shaila Seshia Galvin. I thank Diego Silva and Adil Hasan Khan for their research assistance, and Diego, in particular, for his continued engagement with the project. I am also indebted to the many people at the Graduate Institute who supported this project in one capacity or another. Special thanks to Ghislaine Wharton and Thanh Mai Thi Ngoc for answering numerous questions and for their proficiency in dealing with administrative issues.

The Center for Social Sciences (CSH Delhi), which is part of the French international CNRS research network, was my intellectual home when I was conducting research in India. My deep appreciation goes to my colleagues, in particular Bruno Dorin and Parul Bhandari. I also thank

Yogesh Pai for welcoming me at the Centre for Innovation, Intellectual Property, and Competition (CIIPC) and for offering me the opportunity to teach at National Law University–Delhi.

In Brazil, I wish to thank my colleagues at the Center for Sustainable Development (CDS) and the Professional Master in Sustainability with Traditional Peoples and Territories (MESPT) of the University of Brasília, in particular Mônica Nogueira, Sérgio Sauer, and Gabriela Litre. I am also grateful to Jalcione Almeida, who was the first to welcome me into the Brazilian academic community many years ago; to Patrícia Goulart Bustamante, who has been incredibly welcoming and supportive; and to José Cordeiro de Araújo, who was always open to sharing his knowledge and experience of the Brazilian legislative process.

I owe huge thanks to those who generously helped organize fieldwork trips and interviews. In India, Mallesh K. R. and Suryanarayana Addoor accompanied me to Udupi, Karnataka, and acted as interpreters. In Brazil, Fátima Bresolin and Elisete Kronbauer Hintz went out of their way to organize interviews with soybean producers in Rio Grande do Sul. As Gabriela Pechlaner puts it so aptly, conducting research on ongoing litigation is an exercise in trust, and I sincerely thank all those who agreed to talk to me, often on conditions of anonymity.

I am grateful to the colleagues and friends with whom I have had conversations over the years for their critical engagement and for having been a source of intellectual inspiration. I would like to thank, in particular, Pablo Lapegna, Marc Edelman, Maywa Montenegro, Latha Jishnu, David Jefferson, Matthew Canfield, and Susannah Chapman. My deep gratitude goes to Vinh-Kim Nguyen, for his friendship and mentoring, and for his precious advice over the years; and to Priscilla Claeys, for the friendly complicity, for the encouragement, and for helping me see through some of the issues discussed in *Seed Activism*.

The book was enriched by fruitful exchanges with the fellow seed scholars/activists who participated in the workshop "Seed Activism: Global Perspectives," held at the Graduate Institute in May 2018 with the financial support of the SNSF. The workshop led to a special issue on seed activism, and I thank Jun Borras for supporting its publication in the *Journal of Peasant Studies*.

I presented the material in this book and received input at several academic conferences, including the International Conference of the Commission on Legal Pluralism (2015), the IASC Thematic Conference on Knowledge Commons (2016), the Conference on the Use of Law by Social Movements and Civil Society (2018), the Conference of the BRICS Initiative for Critical Agrarian Studies (2018), and the annual meeting of the American Anthropology Association (2019).

I am indebted to those who have read or reviewed my book proposal and manuscript at various stages of the project. Hannah Gilbert, Hugo Hardy, Marc Edelman, Pablo Lapegna, Priscilla Claeys, Sarah Rugnetta, and Vinh-Kim Nguyen provided feedback and encouragement when this book was still an incipient project. Pablo Lapegna, Diego Silva, Sélim Louafi, and Daniele Manzella offered feedback on specific chapters. I am also indebted to Shalini Randeria for contributing to enrich many of the arguments presented in this book. Special thanks are due to Jack Kloppenburg for generously agreeing to read and comment on the full manuscript. Finally, my appreciation goes to the anonymous reviewers for their insightful and constructive suggestions. Needless to say, I take responsibility for any errors that may remain.

Seed Activism would not be what it is without Tami Parr, who helped me weed through what was originally a much longer manuscript and give the book its actual shape. I thank my mother, Martine Eloy, always my first reader, for her skillful editing; and Keya White and Isabel Wiebe, for lending me a much-appreciated hand with formatting and references. I am also grateful to Laura Portwood-Stacer for her invaluable help in navigating the world of academic publishing.

At MIT Press, Beth Clevenger, and the coeditors of the Food, Health, and the Environment series, Robert Gottlieb and Nevin Cohen, all enthusiastically supported the project. My appreciation goes to the entire team at MIT Press—Anthony Zannino, Marcy Ross, and Kate C. Elwell. I also thank Christine Marra, Mark Woodworth, and Laura Poole for their careful copyediting and proofreading work.

I wish to thank the Swiss National Science Foundation (SNSF) for its generous financial support for the research of this book, as well as for supporting the book's publication in Open Access.

Some of the material in *Seed Activism* has appeared in the following publications and is reproduced with permissions:

- Karine Peschard and Shalini Randeria. 2020. Taking Monsanto to court: Legal activism around intellectual property in Brazil and India. *Journal of Peasant Studies* 47(4): 792–819.
- Karine Peschard and Shalini Randeria. 2019. Regimes proprietários sobre sementes: Contestação judicial no Brasil e na Índia. In *Desenvolvimento e transformações agrárias: BRICS, competição e cooperação no Sul Global*, ed. Sérgio Sauer. São Paulo: Outras Expressões.

I would like to end by expressing my heartful thanks to my family and friends for always being there for me. All my love and gratitude go to my companion, Durval Luiz, for his love and unwavering support; and to my children, Noah and Maya, whose genuine enthusiasm about their mom's publishing a book kept me going in times of doubt.

INTRODUCTION

LAND PILGRIMAGE

An unusual fervor prevails on a late August day in the small town of Cruz Machado, a township in the Paraná state of Southern Brazil. Founded at the turn of the twentieth century by Polish settlers, the township is home to some 18,000 inhabitants, the vast majority of whom live in the countryside. On this day in 2004, more than 20,000 people are expected for the 19th Annual Land Pilgrimage, organized by the Catholic Church's Pastoral Land Commission (CPT).[1] In the early-morning mist, the modern pilgrims descend from buses arriving from all corners of the state. Both a religious and political event, the Land Pilgrimage is organized this year under the theme "I believe in seeds: God's promise, our heritage" (*Creio na Semente: Promessa de Deus, Patrimônio da Gente*).

The opening ceremony is colorful and begins with the blessing of hundreds of farmer-selected seed varieties rescued in the community. The ceremony grows out of the *mística* tradition, a practice of Catholic origin central to Brazilian social movements in which artistic and dramatic performances—singing, dancing, theater, poetry, offerings—are aimed at reasserting certain identities and values as well as a sense of belonging to the movement.[2] While the themes and symbolism in such ceremonies are recurrent, the *místicas* themselves are continually being reinvented.

I.1 "I believe in seeds: God's promise, our heritage." Land Pilgrimage, Paraná, Brazil, 2004

That day, farmers bound by chains march before the silent and attentive crowd, bowed under the weight of bags of seeds bearing the logos of multinational companies that dominate the seed and agrochemical industry—AstraZeneca (Syngenta),[3] BASF, Bayer, DuPont, Monsanto.

In an unambiguous gesture, the farmers then rid themselves of their chains and burdens, and set them on fire. Afterward, the long procession moves on toward a vacant lot outside the town for a day-long celebration of harvest and seeds.

I did not know at the time, but in this vivid staging were all the major themes that were to become central to my research over the following fifteen years: the strengthening of intellectual property (IP) regimes for plant varieties, the consolidation of the global biotech industry, the erosion of agrobiodiversity, and ultimately farmers' dispossession. In addition to the kind of ceremonial resistance I witnessed in Cruz Machado, farmers' unions, civil society activists, and seed companies have increasingly turned to legal resistance, challenging corporate IP rights and practices in court. They question, in particular, the legality of the patents and private IP systems implemented by Monsanto for the collection of royalties on its genetically modified (GM) crop varieties—Roundup Ready soybean in Brazil, and Bt cotton and Bt eggplant in India. As I show, these private IP systems have rendered moot their countries' domestic legislation in the areas of both plant variety protection and farmers' rights. Moreover, these lawsuits have compelled Brazilian as well as Indian courts to grapple, for the first time, with the complex legal questions raised by the extension of IP rights in agriculture in the aftermath of the World Trade Organization (WTO) global trade regime.

In *Seed Activism*, I contend that these lawsuits represent the first significant court challenges to the corporate IP regime in agriculture in the Global South. Notwithstanding the advances and setbacks characteristic of the judicial process, the decisions rendered so far have destabilized the dominant paradigm and hint at the emergence of a new legal common sense concerning both the patentability of plant-related inventions, and the balance between intellectual property, farmers' rights, and the public interest.

SEEDS: BETWEEN ENCLOSURES AND ACTIVISM

Seeds are a fascinating object of study.[4] As the first link in the food chain and the very basis of our food supply, seeds carry tremendous material and symbolic importance, as witnessed by the ancient rituals celebrating sowing and harvesting around the world. Today, seeds straddle the world of the infinitely small—molecular biology and new genetics—and global processes such as transnational peasant mobilizations, global IP and environmental policy, and global capital.

The genetic engineering of seeds in the mid-1980s raised an array of complex issues on virtually every level—scientific, legal, socioeconomic, ecological, health-related, and ethical. In *Seed Activism*, I focus on the proprietary and social justice dimension.[5] Indeed, plant genetic engineering has evolved hand-in-hand with the increasing privatization of seeds and the consolidation of a global seed and agrochemical industry (Howard 2015). In 1996, when GM crops were first commercialized, the top ten companies controlled approximately 40 percent of the global seed market (RAFI 1997). By 2006, that share of the market was controlled by only three companies. The largest company alone, Monsanto, controlled 20 percent (ETC Group 2007). In 2018, following the latest round of megamergers, only four corporations—Bayer-Monsanto, Corteva (formerly DowDuPont), ChemChina-Syngenta, and BASF—controlled more than 60 percent of global proprietary seed[6] sales (40 percent is considered the benchmark of an oligopolistic market) (Howard 2018).[7] During that period, the proprietary seed industry also became intimately linked to the agrochemical industry. This historically unprecedented level of corporate concentration in such a vital sector raises significant concerns over the erosion of agricultural biodiversity, farmers' rights and livelihoods, food security, and, ultimately, the merits of extending IP rights to higher life forms such as plants.

Today, two forms of IP protection coexist for plants: patents and plant breeders' rights (also known as plant variety protection). The two differ in their subject matter, eligibility requirements, and scope. A patent grants an individual the exclusive right to manufacture, use, or sell the product or process on which the patent was granted, usually for a period of twenty years. An invention is eligible for patent protection if it is new, useful, and non-obvious. Under the WTO trade regime, both microorganisms and microbiological processes must be eligible for patent protection.

Plant variety protection was initially developed in the 1940s by a handful of European countries as a form of intellectual protection more appropriate than patents to the nature of agriculture and to the dissemination of new plant varieties. Plant breeders' rights protect a plant variety as a whole and also protect its reproductive material. To be eligible for plant variety protection, a variety must be new, distinct, uniform, and stable (a set of criteria known as DUS).[8] Importantly, plant variety protection, in contrast to patents, include a number of exemptions to the exclusive rights the plant breeder holds. A protected variety can be used either for experimental purposes or as a source of variation to develop new plant varieties (the research and breeding exceptions). Farmers are also allowed to use, save, and exchange a protected plant variety without the authorization of the plant breeder. In industry parlance, this is the "farmers' exception" or "farmers' privilege," but terminology matters greatly, and agrarian movements argue forcefully that seed saving is a *right*—not an exception or a privilege that can be taken away. Moreover, the tendency over time has been to approximate the rights of a plant breeder to those of a patent holder, notably by curtailing the farmers' exception. This is done, for example, by limiting the amount of seeds that can be saved or by restricting the exception to particular crops.

Jack Kloppenburg's seminal work, *First the Seed* (2004 [1988]), inspired a generation of researchers—including myself—to research seeds and plant breeding through the lens of political economy. As Kloppenburg pointed out, for most of agricultural history, seeds have been freely (re)produced and exchanged by farmers. This is because an intrinsic characteristic of the seed—its capacity to reproduce itself—acted as a built-in barrier to capital accumulation. The radical change brought about through agricultural biotechnology—hybridization in the 1930s, and genetic engineering in the 1980s—was to allow capital to overcome social and biological barriers to the capitalization of agriculture by constraining farmers' ability to save seeds.

The first technology developed to commodify the seed—hybridization—consists in cross-pollinating two pure inbred lines. The resulting offspring exhibit enhanced characteristics in terms of size, growth, fertility, and yield compared with the parent lines. Curiously no consensus has been reached concerning the genetic basis of this biological phenomenon, which is called heterosis or hybrid vigor. Hybrid vigor, however, declines in subsequent generations, thus creating an incentive for farmers to buy seeds every year. While some crops like cotton, maize, and rice lend themselves to hybridization, others—for example, soybean—do not. Hybridization thus represents, from a capitalist point of view, an imperfect form of commodification. Hybrids marked the beginning of the seed industry with the introduction of plant breeders' rights.

The second major development in the commodification of seed—plant genetic engineering—can be distinguished from traditional plant breeding by the fact that it operates at the molecular, as opposed to cellular, scale (Krimsky 2019). Genetic engineering made it possible to overcome biological reproductive barriers by using recombinant DNA methods to introduce foreign DNA into the cells of a living organism.[9] Technological developments have played a key role in the commodification of seeds, but only to the extent to which they have been intimately linked to changes in IP regimes (Kloppenburg 2004). With genetic engineering, for the first time, life forms were redefined in patent law as human inventions rather than as products of nature. In turn, the extension of exclusive patent rights to plants spurred the global expansion and consolidation of a proprietary seed industry. Today, agribusiness corporations increasingly use a business model based on licensing fees and royalties rather than on the sale of seeds. Farmers pay a technology fee to corporations, "in effect buying the new genes in a separate transaction from the seed purchase" (Charles 2001, 152). In this way, corporations are licensing genes directly to each farmer.

In the 1980s, developments in genetic engineering intensified the drive toward the patenting of life forms. The decision of the US Supreme Court in the landmark court case *Diamond v. Chakrabarty* (1980) lent judicial support for the patentability of life forms. The patent application for genetically engineered bacteria able to metabolize crude oil was initially denied by the US Patent Office on the basis that living organisms were not patentable. The inventor, a General Electric microbiologist named Ananda Mohan Chakrabarty, appealed, and the US Supreme Court overturned the lower court's decision, ruling that a life form *could* be a human invention as opposed to a product of nature. This represented a real tour de force, considering that patents on life forms were until then thought of as a distortion of IP law that would undermine the patent system (Dutfield 2008).[10] Once it was accepted that a microorganism could in fact be patented, it was a short step to the patenting of more complex life forms, such as plants. This came five years later when another landmark decision by the US Patent Office—*Ex parte Hibberd* (1985)—established the right of plant breeders to obtain protection under the US Patent Act.

Extending patent rights to plants was uncharted territory. There were many gray areas that posed vexed questions. Are plant genes patentable? Can biotech traits be patented as microorganisms? How can one distinguish between a genetic sequence and the plant of which it is part? And, crucially, if an invention is a self-replicating living organism, at what point do the rights of a patent holder become, in patent parlance, "exhausted"? In other words, at what stage of the plant's life cycle does a patent holder lose their exclusive rights?

In the years following the *Ex parte Hibberd* decision, a number of emblematic court cases involving intellectual property in agriculture reached the Supreme Courts of both the United States and Canada. In *Asgrow v. Winterboer* (1995), for example, the US Supreme Court curtailed the scope of the farmer's exception by severely restricting a farmer's right to sell seeds harvested from PBR-protected varieties to neighbors, a common practice known as "brown bag sales." In a legal challenge to the *Hibberd* decision, the US Supreme Court also upheld that sexually reproduced plants were eligible for patent protection (*J.E.M. v. Pioneer Hi-Bred*, 2001). More recently, the US Supreme Court decided that the patent exhaustion doctrine—which states that patent holders lose their exclusive rights to control the use and sale of an article after an authorized sale—does not apply to self-replicating technologies. In other words, a patent owner continues to enjoy exclusive rights over successive generations of seeds, and therefore farmers can be sued for patent infringement if they save seeds from a patented variety (*Bowman v. Monsanto*, 2013).[11]

Around the same period, in neighboring Canada, another lawsuit was drawing the world's attention. Farmer Percy Schmeiser became famous after refusing an out-of-court settlement when Monsanto accused him of having infringed its patent on Roundup Ready canola. Schmeiser argued that he had never sowed Roundup Ready canola and that his fields had been contaminated accidentally (a claim disputed by Monsanto). In 2004, the case went all the way to the Supreme Court of Canada, which ruled in a close five-to-four decision that, no matter how Roundup Ready canola had landed into Schmeiser's fields—whether through genetic contamination or otherwise—Monsanto had a valid patent and therefore owned the genes on Schmeiser's property. However, Schmeiser did not have to pay compensation to Monsanto, as he did not profit from the presence of Roundup Ready canola in his fields. Indeed, he never sprayed Roundup herbicide on his crops and therefore did not take advantage of the Roundup Ready trait.

The judgment was uncompromising in its recognition of patent rights.[12] The Court reasoned that a plant is not a patentable subject matter in Canada, thus limiting the scope of Monsanto's patent to the cells and genes that confer herbicide resistance in canola, and not to the plants themselves. However, the Court proceeded to undermine this argument by holding that infringement occurs when the defendant uses a patented part, even if it is contained in something unpatentable. The justices compared the case to patented Lego blocks assembled in an unpatented structure, a comparison that conveniently obfuscates the fact that Legos are not alive and do not reproduce. The implication is that a patent on a transgenic gene gives the patent owner de facto rights over the plant that

incorporates the said gene. The *Schmeiser* ruling became a landmark decision, not only in Canada but also abroad, where it is often cited in court
cases involving IP and biotech crops (*Monsanto v. Schmeiser*, 2004).

The cumulative result of these court decisions was the consolidation
of an unprecedented proprietary regime in agriculture, which extends
a corporation's IP rights to any seed and plant containing a patented
trait, beyond the first generation and irrespective of how the seed has
been acquired. In both the United States and Canada, corporations' patent
rights on biotech crops are complemented by private contracts signed by
farmers upon the purchase of seeds. This Technology Agreement represents
a significant departure, because it means that farmers no longer *own* the
seeds. Instead, they have a *limited license to use* the seeds purchased, and
they have to commit to using the seeds for a single commercial crop and
not to save or give seeds away. Private surveillance of farmers' fields
and patent infringement lawsuits (or even the mere threat of litigation) act
as further deterrents to seed saving and exchange. In her legal ethnography
of court cases in Canada and the United States, Gabriela Pechlaner (2012)
shows how difficult it is for farmers to contest the loss of control over their
own production. Monsanto has won every single IP lawsuit filed in the
United States and Canada since 1997.[13]

The extension of IP rights over seeds has been likened to a modern form
of *enclosure*, a process through which "things that were formerly thought
of as either common property or as 'uncommodifiable,' or outside the
market altogether, are being covered with new, or newly extended, property rights" (Boyle 2008, 45). Historically, the first enclosure movement
involved the privatization and fencing of formerly common land and
the extinction of customary use rights (Wood 2000). In a similar way, the
extension of IP rights over seeds represents the privatization of seeds as
well as the extinction of ancestral peasant practices of conserving seeds
for next year's harvest, and thus amounts to the expropriation of farmers'
rights over seeds.[14] The legal activism documented in this book represents
a reassertion of the rights that seed enclosures attempt to expropriate.

Indeed, the enclosure of seeds and the related erosion of farmers'
rights and agrobiodiversity have spurred the rise of seed activism during
the past three decades (Fowler and Mooney 1990; Peschard and Randeria 2020). This period coincides with the emergence of transnational

agrarian movements, notably La Via Campesina (Desmarais 2007; Edelman and Borras 2016), and of a new paradigm—food sovereignty, a radical restructuring of our food and seed systems around principles of re-peasantization and re-localization (Claeys 2015). Seed activism therefore encompasses individual and collective actions in defense of both individual and collective rights to seeds. Some forms of seed activism—such as the GM crop-uprooting actions carried out by the French "voluntary reapers" movement (*faucheurs volontaires*) launched in 2003, and the Global March Against Monsanto held throughout the world in 2013–2015—have garnered a great deal of media attention. In contrast, other forms, such as the multiplication of seed saving and sharing networks, are happening quietly under the radar. Contemporary mobilizations around seeds are admittedly extremely diverse in their forms and strategies. Yet it is useful to think about them collectively, because they share a core concern with seed sovereignty—the idea that peasants and farmers must regain control and autonomy over all activities involving seed. Appropriating the biological reproductive potential of plants has been the driving force behind corporate agricultural biotechnologies. Resisting this appropriation via seed saving is therefore central to contemporary struggles over seeds.

THE CORPORATE FOOD REGIME AND ITS DISCONTENTS

The food regime approach first formulated by Harriet Friedmann and Philip McMichael in the late 1980s has gained purchase as an analytical tool to understand the transformation of the global food system historically (Friedmann and McMichael 1989). According to food system scholars, we see, starting in the 1980s, the emergence of a *corporate* food regime.[15] As its name indicates, this regime is characterized by a shift in the organizing principle of the world economy from state to capital (McMichael 2009). This shift is evidenced by the contrast between the Green Revolution of the 1960s and the Gene Revolution of the 1990s. The Green Revolution, founded on high-yield varieties, chemical fertilizers, pesticides, and irrigation, was largely conducted in the international public domain. The Gene Revolution, in contrast, has been driven by the search for private profit in the form of high returns to the shareholders of global corporations in the area of agricultural biotechnology (agbiotech) (Parayil 2003).

Among the distinctive features of the corporate food regime are the reliance on agricultural biotechnology, the privatization of agricultural research, and the creation of global, corporate-friendly IP norms (Pechlaner and Otero 2008; McMichael 2009). Given the neoliberal embracing of IP, it is worth noting that neoliberal intellectuals have historically opposed patents as well as copyrights, which were seen as a "particularly pernicious form of legally-sanctioned monopoly" (Slobodian 2020). In recent decades, neoliberals, notably Chicago School economists, have adopted the utilitarian position that patent rights incentivize innovation. However, the classic rationale behind patents—that they are a necessary incentive to allow firms to undertake risky and lengthy research—rings increasingly hollow (Jaffe and Lerner 2007; Boldrin and Levine 2008). Indeed, evergreening practices, patent thickets,[16] and industry concentration are creating a very real concern that IP is stifling rather than promoting innovation by denying farmers, researchers, and plant breeders access to basic plant materials and processes.

There is also a growing realization that the expansion of corporate IP is negatively affecting farmers' rights and livelihoods and, by extension, the right to food, since small farmers and peasants produce the bulk of food for domestic consumption (Borowiak 2004; Cohen and Ramanna 2007). This has prompted legal scholars to take an interest in the intersection of intellectual property and human rights law, which were previously two distinct legal and policy domains (Helfer and Austin 2011; Helfer 2018).[17] While this literature focuses predominantly on access to medicines and the right to health, a growing wealth of studies examine farmers' rights and the right to food in a human rights perspective (Cohen and Ramanna 2007; Cullet 2007; Haugen 2007, 2020; de Schutter 2009; Santilli 2012; Golay 2017; Bragdon 2020).

An entire body of literature is also devoted to mapping the global trade and environmental regime and its contradictions (Andersen 2008; Tansey and Rajotte 2008; Santilli 2012; Halewood, López Noriega, and Louafi 2013; Shashikant and Meienberg 2015). Indeed, the past three decades have seen a proliferation of new legal instruments governing biodiversity, farmers' rights, and intellectual property. The WTO Agreement on

Trade-Related Aspects of Intellectual Property Rights (TRIPS) obligates member countries to provide patent protection for microorganisms and microbiological processes, and also to give some form of IP protection to plant breeders (WTO 1994). The International Union for the Protection of New Varieties of Plants (UPOV, following its French acronym), is an intergovernmental organization that sets standards for plant breeders' rights. In 1991, UPOV adopted a revised convention (the UPOV 1991 Act), which significantly strengthens the IP rights of commercial plant breeders (UPOV 1991). Finally, the Food and Agriculture Organization (FAO) International Treaty on Plant Genetic Resources for Food and Agriculture (ITPGRFA, hereafter "Plant Treaty") was signed in 2001 (FAO 2001). Its objectives are the conservation and sustainable use of genetic resources for food and agriculture, and the fair and equitable sharing of the benefits arising out of their use.

Some of these instruments, namely, the TRIPS Agreement and the UPOV Convention, have been primarily driven by the corporate interests of the pharmaceutical and seed industry (Sell 2003); others, such as the FAO Plant Treaty, have been motivated by the protection of agrobiodiversity and farmers' rights. It is therefore not surprising that these legal instruments are riddled with conflicts. These contradictions are replicated at the domestic level, since different ministries typically develop overlapping legislation that cover the same subject matter but operate on divergent principles and respond to different constituencies (Newell 2008). In India, for instance, the Ministry of Agriculture oversaw the Protection of Plant Varieties and Farmers' Rights Act (PPVFR Act), the Ministry of the Environment oversaw the Biological Diversity Act, and the Ministry of Science and Technology oversaw the Patent Act. Bt cotton varieties come under all three types of legislation: the biotech trait can be patented; the plant variety containing the biotech trait can be protected by plant breeders' rights; and, as living modified organisms, Bt cotton varieties are subject to biodiversity and biosafety regulations.

While considerable research has been done on the emerging global IP regime for plant resources, more empirical research is needed on the concrete forms that these global norms take on the ground.[18] To paraphrase Stephen Brush (2013), we should be less concerned with presenting the intricacies of national and supranational laws and agreements, and more focused on how these legal regimes actually play out. Doing so illuminates the real difficulty of implementing international regimes governing access to plant genetic resources at the national level, as revealed by the case of Bt brinjal (see Appendix C). It also shows how corporations use "private orderings" (in this case, licensing contracts) to bypass public law. In spite of the expansion and strengthening of global IP norms, states have considerable room to maneuver, and important differences remain in domestic legislation pertaining to intellectual property and agriculture. Add to this different patent cultures and agri/cultures, and the result is a wide array of *actually existing* IP regimes.

In addition to the local expressions of global IP norms, it is important to pay attention to the growing contestations surrounding the IP dimension of the corporate food regime, notably in the courts. Resistance in the legal forum is key here, because courts are the final arbiters of disputes over the appropriate scope of patents, and their interpretations are binding on patent offices (Dutfield 2006). *Seed Activism* is part of the growing trend, among food systems scholars, to study not only the regimes themselves but also, and importantly, the transitions between them, in particular the role of social movements in hastening change (McMichael 2009, 2013). This book prompts a larger question: what potential does legal activism around biotech crops hold for the unfolding of the current corporate food regime? Do these legal developments merely signal a fragile departure and a vulnerable resistance?[19] Or do they, in their manifold and sometimes unintended ramifications, have the potential to influence or even derail the dominant corporate IP paradigm at the core of the corporate food regime? It is too early to give a definite answer to this question, but my contention is that these lawsuits have evolved to challenge fundamental dimensions of the corporate food regime in agriculture, including the primacy of the commercial rights of patent holders over the fundamental rights of farmers, and the very patentability of genes and plants.

LEGAL CONTROVERSIES OVER INTELLECTUAL PROPERTY AND BIOTECH SEEDS IN BRAZIL AND INDIA

As Monsanto, with the backing of the US government, strove to export the proprietary regime it had developed in the United States to other major GM crop producers in the early 2000s, the locus of social and legal conflicts over intellectual property and biotech crops shifted to the Global South, notably Argentina, Brazil, and Colombia, and on the other side of the globe, India and Pakistan.[20] Court cases in North America arising from the new legal regimes in agriculture have been the object of considerable academic interest (see, for example, Ewens 2000; Kloppenburg 2004; Müller 2006; Aoki 2008; and Pechlaner 2012), however, in-depth analyses of similar developments in the Global South are few and far between.[21] Given that the agricultural and legal landscapes in these countries differ significantly from those of the United States, could these lawsuits around intellectual property and agricultural biotechnology have a different outcome than in North America? This book endeavors to answer this question, based on the experience of Brazil and India.

Brazil and India offer rich grounds for a comparative exploration of these issues. The two countries are, with Argentina, the top GM-crop producing nations in the Global South.[22] Brazil ranked as the second top producer of biotech crops in 2019 with 53 million hectares, while India ranked fifth with 12 million hectares (ISAAA 2019).[23] Their agrarian structures, however, differ significantly. In Brazil, a large and powerful export-oriented agribusiness sector coexist with a family farming sector that plays a vital role in food production and food security.[24] India's agriculture, by contrast, is dominated by large numbers of small and marginal farmers.[25] In recent decades, several factors—including fragmented landholdings, lack of infrastructure, volatile prices, dependence on middlemen, and high indebtedness—have combined to cause persistent agrarian distress in India.

Brazil and India are both biologically "megadiverse" countries, a term used to refer to a group of seventeen countries that are located in subtropical and tropical regions and that harbor the majority of the Earth's species. With 15 percent to 20 percent of the world's biological diversity, Brazil is considered the most biologically diverse country in the world (CBD n.d. [1]). It is the center of origin and diversity for a number of cultivated plants, such as manioc and peanut, and is home to at least 43,020 known plant species. India is the center of origin and diversity for a large number of food crops, notably rice, and is home to some 45,500 documented species of plants (CBD n.d. [2]). Brazil and India are also important repositories of traditional knowledge associated with biological diversity. In addition, both countries maintain a strong tradition of publicly funded agricultural research[26] and are home to some of the most important national plant germplasm collections worldwide.[27]

As biologically megadiverse countries and large agricultural producers, Brazil and India are key players in the contentious global negotiations over agricultural trade and genetic resources. The two countries have been actively involved in international negotiations over the WTO TRIPS Agreement, the United Nations Convention on Biological Diversity (CBD), the FAO Plant Treaty, and the Nagoya Protocol on Access and Benefit-sharing (United Nations 1992; FAO 2001).[28] In the past two decades, Brazil and India have also increased their cooperation through forums such as IBSA (India–Brazil–South Africa) and BRICS (Brazil–Russia–India–China–South Africa), a relationship strained by the election of right-wing populist governments in the mid-to late 2010s.

Importantly for the purposes of this book, Brazil and India are home to major, precedent-setting court cases involving intellectual property and biotech crops. In 2009, in the small town of Passo Fundo, in southern Brazil, a farmers' union filed a class action against Monsanto concerning royalties on genetically modified Roundup Ready soybeans. Rural unions questioned Monsanto's practice of charging royalties on harvested soybeans, as opposed to the conventional practice of charging royalties on the sale of seeds. Charging royalties on harvested soybeans represented a major change, since it extended Monsanto's rights to a farmer's production, thereby effectively doing away with a farmer's right to freely save seeds for replanting. As it made its way through the judicial system over the course of the next decade, the class action became a precedent-setting, multibillion-dollar lawsuit concerning some five million Brazilian farmers. In the course of this legal battle, the farmers' unions obtained favorable rulings and also suffered setbacks. Further, they uncovered disturbing facts about Monsanto's Brazilian patents.

Around the same period, in Mattu, a small village on the coast of Karnataka in southern India, farmers learned that a local eggplant variety in cultivation for four centuries had been used to develop Bt brinjal, a genetically modified eggplant, without their knowledge or consent. Mattu Gulla, as the variety is known, is one of the local eggplant varieties at the center of public interest litigation and criminal prosecution on grounds of "biopiracy," the misappropriation of plant genetic resources. Indeed, in 2010, an Indian nongovernmental organization (NGO) called Environment Support Group drew public attention to the fact that the international public–private consortium that developed Bt brinjal had not obtained permission from Indian biodiversity authorities to access local eggplant varieties, as required under the Indian Biological Diversity Act (GoI 2002a).

The Indian government and national seed companies also engaged in a tug-of-war with Monsanto over the regulation of royalties for genetically modified Bt cotton, the subject of multiple lawsuits and complaints in the Delhi High Court and in the Competition Commission of India (CCI). Bt cotton is genetically modified to produce a protein that behaves as a toxin against various bollworms. This case stands out as the first attempt by a government to intervene and regulate royalties on genetically modified crops. One lawsuit filed by Monsanto against an Indian seed company for patent infringement led the Delhi High Court to revoke Monsanto's Indian patent on Bt cotton in April 2018, in the first decision to examine the legality of patents on biotech traits in India. This decision was subsequently suspended by the Supreme Court, which instructed the Delhi High Court to conduct a full trial.

Taken together, these three legal disputes shed light on the contested nature of IP regimes for biotech crops in two major GM crop producers in the Global South. While the legal cases arising in North America have been the object of extensive analyses by legal scholars and social scientists, the more recent court challenges involving intellectual property and plant genetic resources in the Global South are much less documented. In contrast to the controversy over the environmental and health dimension of Bt brinjal, the IP dimension of the case has attracted relatively little coverage and analysis (Abdelgawad 2012). Likewise, no in-depth analysis has been done of the ongoing struggle over royalties in Brazil, despite the fact that it is widely regarded as one of the most important cases worldwide involving intellectual property in agriculture. Felipe Filomeno's fine-grained analysis of the implementation of royalty collection systems in South America includes Brazil, but the court cases were incipient at the time he conducted his research (Filomeno 2014). As for the Bt cotton trait fee controversy, it has recently begun to attract more attention among legal scholars and social scientists (Agarwal and Barooah n.d.; Manjunatha et al. 2015; Sathyarajan and Pisupati 2017; Stańczak 2017; Van Dycke and Van Overwalle 2017).

While this book was not initially intended to be about Monsanto per se, the fact that Monsanto stands at the center of these legal disputes is no coincidence either. Ever since the commercial introduction of genetically modified organisms (GMOs) in agriculture in 1996, Monsanto has aggressively pursued its commercial interests both in the United States and outside of it. Monsanto filed for patent protection for its GM traits in various jurisdictions, but also devised and implemented unprecedented systems for the collection of royalties and the surveillance of farmers. It has not shied away from suing farmers, including its own customers, for patent infringement.[29] As the lawyer in the RR soybean class action observes, although other companies followed suit and adopted similar IP models and practices, it is Monsanto that has been the architect of the royalty collection system in Brazil (Interview #29B). In 2018, the German chemical giant Bayer acquired Monsanto for 66 billion USD, but I refer to "Monsanto" throughout this book since the events described here took place prior to the acquisition.

Building on these legal cases, I set out to answer three interrelated sets of questions related to the role of corporations, states, and civil societies in the overhaul of IP regimes in agriculture. First, what legal tactics have Global North corporations developed to assert IP rights in countries with stronger farmers' rights legislation and, conversely, more limited IP protection? What has been the role of the state in the implementation of IP regimes for biotech crops in these countries? Finally, how are agbiotech patents and royalty collection systems contested in the courts? By which actors, and on what grounds?

In *Seed Activism*, I demonstrate that Monsanto designed and implemented private royalty collection systems adapted to the specificities of each country's crops and agrarian conditions. As I will show, these systems succeeded in ensuring that Monsanto would enjoy the same extraordinary degree of IP rights in Brazil and India as it does in the United States, irrespective of the fact that the patent and plant variety protection laws of these countries differ significantly from those of the United States.

Second, I show that states in the Global South have been complicit in the implementation of these private biotech IP regimes. In transposing the WTO TRIPS norms into their domestic legislation, neither Brazil nor India took full advantage of the flexibilities available to them. What is more, corporations operating in those countries were largely given a free hand to implement private royalty collection systems that maximized their profits at the expense of farmers.

Third, I argue that court challenges to patents and royalty collection systems have created unexpected issue-based, short-term alliances among actors with varied political agendas pursuing different long-term goals (Peschard and Randeria 2020). For instance, family farmers and large landowners in Brazil have come together to contest Monsanto's monopoly claims in national courts. Likewise, in India, seed sovereignty activists as well as Hindu ultranationalists have found themselves on the same side in the judicial battle against the corporation. I examine the strengths and ambiguities of such fleeting alliances while delineating the power dynamics shaping legal activism around IP and biotech crops.

Seed Activism thus sheds light on the role of biotech corporations in the contemporary food regime, on the transposition of supranational norms at the domestic level, and on the nature and prospects of legal activism in the Global South. While Brazil and India are not representative of the whole of the Global South, their relatively privileged economic and political positions mean that judicial developments there can cause ripple effects elsewhere.

BEHIND THE BOOK

As an anthropologist studying legal processes and mobilizations, I subscribe to an ethnographic approach that follows legal processes as they unfold in people's daily lives—an approach sometimes referred to as "law in action," as opposed to "law on the books." The basic tenets of this approach are that law must be studied from the bottom up and in context. Such an approach entails engaging, as much as possible, with the full range of those directly or indirectly involved in these legal cases. It also means paying particular attention to the role of power—material, institutional, and discursive—in shaping law and the legal system. Only by doing so can we begin to understand the variety of ways in which people resist, accommodate to, and ultimately shape new legal regimes in agriculture. Such an approach also enables us to grasp the wider repercussions of legal controversies and litigation beyond formal court decisions. Research for this book has thus taken me from the Geneva headquarters of UPOV to the soybean fields of Southern Brazil; and from the New Delhi offices of corporate lawyers to a Hindu temple where the Mattu Gulla eggplant has been at the center of a centuries-old religious celebration.

I conducted the bulk of the research between October 2015 and September 2019, a period during which I lived in India (2015–2016) and Brazil (2016–2019), but the book also builds on long-term experience researching seed-related issues, in Brazil since 2004 and in India since 2012. I rely on a combination of in-depth interviews, participant observation, and legal case analysis. In total, I conducted ninety open-ended interviews with the various parties involved in litigation—farmers, rural union leaders, expert witnesses, government officials, industry spokespeople, plant scientists, NGO workers, civil society activists, legal researchers, lawyers representing both sides in the disputes, as well as a judge (see Log of Interviews, Appendix B). I also studied a large body of legal documents—including sublicensing agreements, expert witnesses' report, court documents, and judicial decisions—and attended public hearings, court sessions, and parliamentary commissions on related issues. In addition, I relied on an extensive search of the newspaper record to reconstruct the chronology of events in these legal controversies.

Writing about unfolding events is both a curse and a blessing: along with the thrill of treading new ground comes the challenge of trying to analyze a situation that is constantly evolving before one's eyes. The Bt cotton trait fee dispute, for example, was not originally part of the project. But when it erupted into a full-blown controversy on the national scene in late 2015, shortly after I began my research, its relevance and the parallels with the Roundup Ready soybean class action were too good to be ignored.

The challenge of accompanying unfolding events is compounded by the unpredictability of the judicial system. For instance, while I was initially hopeful that the Supreme Court of India would deliver a decision in the Bt brinjal case, this hope faded as the realization sank in that this was not simply a case of judicial backlog and that the case might be indefinitely in limbo for political reasons. While these delays and setbacks can be as frustrating for researchers as they are for litigants, they speak to the power relationships and interests involved in these controversies. In the three lawsuits, the parties have not yet exhausted all of the legal remedies available to them. This being said, the lawsuits have come a long way since they were originally filed, and they offer ample material for analysis.

A great deal has changed since I embarked on this project in 2015. That year, Monsanto's patent on the Roundup Ready herbicide-tolerance trait became the first major agbiotech patent to expire and enter the public domain, prompting widespread speculation around a "post-patent" era. Between 2015 and 2018, the number of corporations controlling the global seed and agrochemical market went from six to three, heightening concern over the oligopolistic nature of the seed market (Clapp 2021).[30] In 2020 and well into 2021, India was rattled by historical farmers' protests over the passing of legislation dismantling rules for the sale, pricing, and storage of farm produce, which farmers feared would leave them at the mercy of transnational corporations. As I write these lines in September 2021, the increasingly viable threat of our global food chain collapsing under the combined pressure of climate change and the COVID-19 pandemic makes understanding the proprietary dimension of our food regime more urgent than ever.

1

BRAZIL, INDIA, AND INTELLECTUAL PROPERTY IN AGRICULTURE

At long last, "everything under the sun made by man" . . . is potentially patentable.
—Comment on the *Ex parte Hibberd* ruling published in *Nature* (Van Brunt 1985)

Following the development of plant genetic engineering in the 1980s, the United States and other industrialized countries extended patent rights to genetically engineered processes and products. In the 1990s, the United States and its pharmaceutical and agbiotech industries then worked to spread US intellectual property (IP) standards globally.[1] During the Uruguay Round (1986–1994) of multilateral trade negotiations, the United States spearheaded efforts to include intellectual property on the agenda of the General Agreement on Tariffs and Trade (GATT), along with the support of other industrialized countries, including Japan, Australia, Canada, New Zealand, Switzerland, and the European Community. These countries had strong pharmaceutical and agrochemical industries that held the vast majority of patents in force around the world. In 1995, the World Trade Organization (WTO) replaced the GATT. One of WTO's founding agreements, the Agreement on Trade Related Aspects of Intellectual Property Rights (TRIPS), introduced global IP norms in the area of plant biotechnology and plant varieties.[2]

What was at stake was not so much the adjustment of IP regimes as the extension of US and European standards of intellectual property—a distinctly Western tradition—to the rest of the world. Prior to the TRIPS Agreement, a majority of countries expressly excluded plant varieties from IP protection. This was especially the case for countries in the Global South, but even for some countries with industrialized agricultural sectors.[3] Indeed, the international plant-breeding system was based on the premise that everyone would benefit from free access to and wider exchange of plant genetic material and knowledge.

Countries in the Global South, notably Brazil and India, were aware that, as a negotiator himself admitted, "they had nothing to gain but much to lose" from taking on new commitments in the area of intellectual property (Ganesan 2015, 213).[4] Their initial position was that IP was best dealt with by the World Intellectual Property Organization (WIPO), and that existing IP conventions were adequate. Around 1988–1989, Brazil and India moved from "staunch opposition" to the inclusion of higher standards of IP protection in the GATT mandate to "hesitant acceptance" (Ganesan 2015; Tarragô 2015). A number of factors explain this turnabout. First, during that period, both countries saw the coming-to-power of governments favorable to market-friendly policies, privatization, and foreign investment. In Brazil, the return to democracy in the late 1980s was accompanied by a shift to market-oriented policies. In India, the New Economic Policy of 1991 marked the turn toward neoliberal policies. Second, both countries were under significant economic pressure from the United States. The latter had previously sanctioned Brazil for its lack of patent protection for pharmaceuticals and agrochemicals and had placed India on the newly created US Special 301 Watch List due to its lack of pharmaceutical patent protection.[5] Finally, it seemed that both Brazil and India were finding it difficult to withstand the unified offensive as well as the robust lobbying of big corporations and their governments on this front, owing to "the absence of a unified position among the developing countries, which had either little expertise in the subject matter or limited capacity to resist the pressures" (Tarragô 2015). Ultimately, Brazil's and India's changes of heart were in fact pragmatic decisions born out of the realizations that the United States would not budge on this issue and that intellectual property could be used strategically as a bargaining chip to obtain gains on other issues, such as enhancing market access for agricultural, textile, and tropical products (Ganesan 2015; Tarragô 2015).

Both Brazil and India started out negotiations of the TRIPS Agreement from a defensive position. On the scope of patentable subject matter, Brazil stated that patents should be granted to inventions fulfilling the criteria of patentability, with the exception of inventions that are contrary to morality, religion, public order, or public health, while at the same time bearing in mind the public interest as well as technological and economic development (GATT 1989b). India went further, and argued that:

Every country should be free to determine both the general categories as well as the specific products or sectors that it wishes to exclude from patentability under its national law taking into consideration its own socio-economic, developmental, technological and public interest needs. It would not be rational to stipulate any uniform criteria for non-patentable inventions applicable alike both to industrialised and developing countries or to restrict the freedom of developing countries to exclude any specific sector or product from patentability. Developing countries should be free to provide for process patents only in sectors of critical importance to them such as food, pharmaceutical and chemical sectors. (GATT 1989a)

In contrast, the United States defended stronger IP rights in all fields of technologies, including longer terms of patent protection, stiffer enforcement measures, short transition periods for implementation, and limited ability for governments to set conditions to grant those rights.

The question of the patentability of living material, including plants, rapidly became one of the main bones of contention during the negotiations of the TRIPS Agreement. At one end of the spectrum, the United States demanded no exceptions from the general rule of patentability (GATT 1987). At the other end, a group of countries from the Global South that included Brazil and India—known as the Group of 14—suggested excluding from patentability inventions contrary to public health; discoveries; material or substances already existing in nature; methods for the medical treatment of humans or animals; and nuclear and fissionable material (GATT 1990). The European Union's position offered a middle ground, with exclusions from patentability limited to plant and animal varieties; processes that were essentially biological; and inventions contrary to public order or morality (GATT 1988).

The outcome of negotiations over these issues was a slightly modified version of the EU proposal. Article 27.3(b) states that members may exclude from patentability: "Plants and animals other than micro-organisms, and essentially biological processes for the production of plants or animals other than non-biological and microbiological processes. However, Members shall provide for the protection of plant varieties either by patents or by an effective sui generis system or by any combination thereof" (WTO 1994). In other words, that article requires member countries to extend patents to microorganisms and microbiological processes, and to provide some form of IP protection for plant varieties.

Some government officials, however, were cognizant of the ambiguities surrounding certain provisions of the TRIPS Agreement, notably Article 27.3(b). As the Indian Minister for Commerce and Industry put it at the time, "We are all aware that the text of the TRIPS is a masterpiece of ambiguity, couched in the language of diplomatic compromise, resulting in a verbal tight-rope walk, with a prose remarkably elastic and capable of being stretched all the way to Geneva" (Ministry of Commerce and Industry 2002).

First, no single accepted definition of "microorganism" exists, and some argue that the term itself is inherently flawed (Singh 2015). Moreover, genetic engineering is not about microorganisms as much as it is about genetic sequences and genes; the term "microorganism" seldom appears in plant-related patent applications. Patent offices worldwide have been left to deal with the resulting ambiguity. Second, the meaning of "an effective sui generis system for plant varieties" remains to this day open to interpretation. *Sui generis* means, literally, "of its own kind." The phrase implies that countries can develop legislation adapted to their needs and objectives as long as they respect the minimum IP standards established in Article 27.3(b), but there is still no agreed-on understanding of that phrase.

As one Indian negotiator concluded, "A general lack of understanding of all the issues involved and the broad wording of the provisions helped limit contentious negotiations" and also allayed concerns around the patenting of microorganisms and sui generis protection for plant varieties during the TRIPS negotiations (Ganesan 2015, 230). However, as his Brazilian counterpart observes, in the end, TRIPS negotiators were not able to resolve the quandary of the extension of patents to living materials in a manner that could assuage the concerns of farmers and traditional knowledge holders (Tarragô 2015, 246). These ambiguities, as we will see, were soon to resurface as WTO member countries moved to pass legislation implementing the TRIPS Agreement domestically.

Because Article 27.3(b) and its extension of intellectual property to plant biotechnology and plant varieties was contentious, one of the compromises made during the negotiations was the inclusion of a mandatory review of this provision within four years of the agreement's entry into force. An objective of the review was to clarify what constituted an effective sui generis system (WTO n.d.). The review procedure, however, was never implemented, despite repeated demands from countries in the Global South, including India, Brazil, and the African Group. Indeed, the United States and the European Union consistently blocked the review, making the political calculus that it was preferable to stick to the current wording of Article 27.3(b) than to risk reopening a Pandora's box.

Position papers submitted during the debates surrounding the proposed review of Article 27.3(b) show how differences endured even after the entry into force of the TRIPS Agreement. In a 1999 submission, the Indian delegation called for a substantive review of Article 27.3(b), including (1) whether and what form of exclusion from patentability should apply to plants and animals; (2) the effect of protection granted to microorganisms as well as to both nonbiological and microbiological processes; and (3) the sui generis system and its effectiveness (WTO 1999). The Indian delegation raised the ethical issue of the extent to which private ownership could extend to life forms and of "the appropriateness of the concept of intellectual property, as it was understood in the industrialized world, in the face of the wider dimension of rights to knowledge, their ownership, use, transfer and dissemination" (WTO 1999, 11). India went further, recommending that "patents should be excluded for all life forms or, if that were not possible, at least for those based on traditional or indigenous knowledge and essentially derived products and processes" (WTO 1999, 11). On the issue of microorganisms, the Indian delegation reiterated the importance of distinguishing between inventions, which could be patented, and discoveries, which could not. Moreover, while the Indian delegation accepted that a man-made microorganism (for example, a genetically engineered bacterium) met the requirements of patentability, "it questioned whether patents could extend to cell-lines, enzymes, plasmids, cosmids,[6] and genes" (WTO 1999, 12). The Indian delegation noted that these issues had not been fully explored during the negotiations and that it contained terms on which even scientists could not agree. These issues, that delegation argued, should not be left open to future interpretation by the technical panels of patent offices. Finally, it added that even microorganisms could be excluded from patentability for the reasons stipulated in Article 27.2, namely "to protect *ordre public* or morality, including to protect human, animal or plant life or health, or to avoid serious prejudice to the environment" (WTO 1994). Finally, India's delegation pointed out that adhering to the International Union for the Protection of New Varieties of Plants (UPOV) was only one option among others to meet the TRIPS obligations and insisted that countries had "great latitude to develop an effective means of protection," taking into account their "own public policy objectives, including developmental and technological objectives" as well as their obligations undertaken in the context of other international agreements.

The Brazilian delegation positioned itself in a middle ground between the US position of "extending patentability to all life forms" and the Indian position of "excluding patentability for all life forms," favoring instead maintaining "the status quo of Article 27.3(b) as it was" (WTO 1999, 24). While Brazil had chosen to adhere to UPOV, its delegation noted that it was important to consider the national experiences of other countries in developing sui generis systems. Finally, the Brazilian delegation supported the review of Article 27.3(b), particularly on the issue of the protection of traditional knowledge.

The wording of Article 27.3(b) gave countries considerable flexibility to develop a sui generis system for plant varieties. Indeed, while the TRIPS Agreement stipulates that countries must provide some form of IP protection for plant varieties, it leaves considerable leeway as to how this is done. However, industrialized countries, whose seed industry stood to benefit from strong plant breeders' rights, skillfully maneuvered to get countries to meet the TRIPS requirements in matters of intellectual property for plants by adhering to UPOV.[7]

UPOV is an intergovernmental organization that enforces intellectual property on plant varieties, more specifically plant breeders' rights. UPOV was established in 1961 by a handful of European countries with a strong plant breeding industry. As of 2020, countries that were members of UPOV were signatory to either the 1978 or the 1991 Act of the UPOV Convention. The main difference between the two versions of the convention is that plant breeders' rights are significantly reinforced under the 1991 Act. Under this act, for example, seed saving becomes an optional exemption for countries; seed saving is restricted to a farmer's own use, and "must safeguard the legitimate interests of the breeder." Membership in UPOV is entirely voluntary, but countries have been under pressure to meet their TRIPS obligations by adhering to UPOV. However, by joining UPOV, countries effectively gave up the possibility of developing a sui generis legislation adapted to their needs and interests, and at the same time implemented stronger IP rights than required by the TRIPS Agreement (known as "TRIPS-plus" provisions). As we will see in the remainder of this chapter, this was the path taken by Brazil but not by India.

INCORPORATING ARTICLE 27.3(B) INTO DOMESTIC LAW: THE CASE OF BRAZIL

Prior to its entry into the World Trade Organization, the laws of Brazil did not provide IP protection for plant varieties. A Seed Bill enacted in 1965 established norms for seed production and trade, but there was no specific protection for breeders' rights. At the time, the public sector played an important role in plant breeding, and the new cultivars it developed remained in the public domain. Private seed companies multiplied and distributed seeds and were active in breeding crops that were amenable to hybridization, such as corn. Following the entry into force of the TRIPS Agreement, Brazil overhauled its legislation on plant varieties by amending the Industrial Property Act and by passing the Plant Variety Protection Act.

Brazil's post-TRIPS legislative overhaul was not the country's first attempt at introducing breeders' rights legislation.[8] In the mid-1970s,

International Plant Breeders, a group owned by the Dutch-British conglomerate Royal Dutch Shell, which controlled the largest share of seed sales worldwide at the time, worked closely with the Brazilian Seed Producers Association (ABRASEM), the country's Ministry of Agriculture, and the Brazilian Agricultural Research Corporation (EMBRAPA) to introduce plant variety protection legislation (Pelaez and Schmidt 2000). However, their collective efforts collapsed when a preliminary version of the bill and regulations was leaked to the public in February 1977. The document immediately caused public outrage. The legislature of the State of São Paulo, for example, passed a motion condemning the bill as an attempt to denationalize the Brazilian seed production sector for the benefit of foreign companies, as well as a threat to Brazilian farmers, who would have to pay higher prices for seeds. The motion stated that "protecting seeds through patenting amounts to protecting the commercial interests of large economic groups in already developed countries, to the detriment of the real interests of our farmers and of national private enterprises" (Paschoal 1986, xv).

The campaign against the bill garnered support in the following months, including that of the influential Brazilian Society for the Progress of Science (SBPC), several professional associations of agronomists, and a number of members of Congress. Headlines of major newspapers reflected the growing negative public sentiment: "Agronomists Against Seed Project" (*O Estado de São Paulo*); "The Seed Is Ours" (*Veja*); and "Seeds: Multinational Control?" (*Diário do Comércio e Indústria*) (Paschoal 1986). Mobilization was fueled by fears of multinational corporations' gaining a monopoly over seeds and of potential price rises if seeds were to come under the control of the private sector, and had strong nationalist overtones. Another important factor that contributed to the bill's disrepute was the role that IPB (a foreign lobby) and Agroceres (a private seed company) played in the drafting of the bill. Under pressure, the Ministry of Agriculture announced the bill's indefinite deferment in August 1977 (IPB closed its Brazilian office shortly thereafter).[9]

Twenty years went by before plant variety protection issues were back on the table in Brazil. Much had changed in the two decades since the first unsuccessful attempt to introduce plant variety protection legislation in

1977. The nation's military dictatorship and its national-developmental ideology had been replaced by a democratically elected, market-oriented government. Transnational seed companies had entered the Brazilian market (not to mention Brazilian politics) and were pressuring for strong IP rights as transgenic varieties were about to be introduced. Finally, the international context had also changed; Brazil had given in to international pressures and signed onto the TRIPS Agreement and UPOV. According to a legislative consultant, what eventually became known as the Plant Variety Protection Act, or the PVP Act, "was passed under international pressure and, internally, under pressure from agribusiness, that is, large seed producers and multinational corporations entering Brazil at that time in the area of biotechnology and transgenic crops. They were the ones who needed the [PVP Act]" (cited in Peschard 2010).

Among the groups pushing for new legislation were ABRASEM, the National Agricultural Confederation (CNA), EMBRAPA, and the companies that were soon to set up the Brazilian Plant Breeders Association (BRASPOV). Opposition to the PVP Act was led this time by a caucus of the left-wing Workers' Party (PT), jointly with civil society organizations such as the Brazilian Institute for Consumer Protection (IDEC), an agroecology and family agriculture NGO (AS-PTA), and Greenpeace-Brazil.[10] As we can see, the locus of opposition to the bill had shifted from within the professional body of agronomists and elected officials (deputies, senators, and politicians) to civil society. Civil society organizations were able to limit the scope of certain articles of the bill but not prevent its adoption.

The PVP Act was passed in 1997. The Act established a significant legal benchmark, since it introduced IP rights for plant varieties in Brazil. The PVP Act states that "The protection of intellectual property rights in plant varieties is effected through the grant of a Plant Variety Protection Certificate, which shall be considered a commodity for all legal purposes and the sole form of protection in the Country for plant varieties and the rights therein that may be invoked against the free use of sexually or vegetatively propagated plants or parts thereof" (RFB 1997, Art. 2).

The PVP Act is based on the 1978 Act of the UPOV Convention and was drafted with the explicit objective of adhering to UPOV (Araújo 2010). Initial drafts of the bill did not contemplate farmers' rights to freely save, exchange, and sell seeds, and the limited farmers' rights provisions that were eventually included in the PVP Act and the revised Seed Act were only secured after an arduous struggle by civil society organizations. This includes Article 10 of the PVP Act, which recognizes the right of farmers to keep and plant seeds from protected varieties for their own use. According to that article, farmers may store and plant seeds for their own use, and they may also use or sell the product of their plants as food or raw material (except for reproductive purposes). An exception is made for small rural producers, who can also multiply seeds from protected varieties to give away or exchange, but only in dealings exclusively with other small rural producers (RFB 1997).[11] In 1999, Brazil joined UPOV just before the doors closed to the 1978 Act. By doing so, the country avoided being forced to join the more restrictive 1991 Act.[12]

To fulfill its new obligations under Article 27.3(b) of the TRIPS Agreement, Brazil also had to revise its patent legislation. A bill had been tabled in 1991 by President Fernando Collor de Mello, the first democratically elected president after the end of the military dictatorship.[13] Collor de Mello was elected on a neoliberal platform, and the bill was intended as a clear gesture to the international community, in particular the United States, that Brazil was adhering to the new rules governing IP rights and trade and was thus a reliable trading partner (de Alencar and van der Ree 1996). Civil society mobilized against the bill, in particular against the fact that it allowed the patenting of life forms, thus delaying its adoption by five years. The new Industrial Property Act (as the Patent Act is called in Brazil) was sanctioned by President Fernando Henrique Cardoso in 1996.[14]

The Industrial Property Act of 1996 allowed for the patenting of life forms for the first time. Under the law, to be patentable, an invention must meet the standard requirements of novelty, inventive activity, and industrial application (Art. 8). This excludes "all or part of natural living beings and biological materials found in nature, even if isolated therefrom, including the genome or germoplasm of any natural living being, and natural biological processes," which are not considered as qualifying as inventions (RFB 1996, Art. 10, IX). The Act further specifies that transgenic microorganisms that satisfy the three requirements of patentability—novelty, inventive activity, and industrial application—are patentable but that all or part of living beings, and mere discoveries, are not (RFB 1996,

Art. 18). An explanatory paragraph defines transgenic microorganisms as "organisms, except for all or part of plants or animals, that express, by means of direct human intervention in their genetic composition, a characteristic normally not attainable by the species under natural conditions" (RFB 1996, Art. 18). In sum, both microorganisms and microbiological processes are eligible for patent protection, as required by Article 27.3(b), but not animals, plants, plant parts, or plant varieties.

Under industry pressure, the Industrial Property Act also included a transitional provision known as the pipeline mechanism. That mechanism allowed companies to apply for patents on products or processes invented before the Act came into force and patented abroad, provided they had not yet been marketed in Brazil. As a TRIPS-plus provision—one that goes beyond the minimum requirements of the TRIPS Agreement—the pipeline mechanism came under severe criticism throughout civil society. As we will see in the next chapter, their concerns were not unfounded. More than 1,100 patents on biotechnological inventions in the pharmaceutical and agricultural sectors—including patent applications related to Roundup Ready soybean—were granted in Brazil between May 1996 and May 1997, under the pipeline mechanism (Muniz 2018). Roundup Ready soybean is genetically engineered to resist the direct application of Monsanto's Roundup herbicide and was the first GM crop variety cultivated in Brazil.

INCORPORATING ARTICLE 27.3(B) INTO DOMESTIC LAW: THE CASE OF INDIA

Like Brazil, India did not provide intellectual property protection for plant varieties prior to joining the WTO. The first draft of its plant variety protection bill, introduced in 1993, met with considerable opposition and prompted mass demonstrations by farmers, termed *beej satyagraha*, or "seed protest" (Seshia 2002). As a result, revised drafts were introduced in 1997, 1999, and 2000. From January to August 2000, a Joint Parliamentary Committee held public hearings throughout India. After a seven-year struggle and no fewer than five drafts, the Protection of Plant Varieties and Farmers' Rights (PPVFR) Act was finally passed in 2001 (GoI 2001).

India stands out as one of the few countries worldwide to have introduced sui generis legislation in the area of plant variety protection. Its

PPVFR Act provides standard IP protection for plant breeders but seeks to balance their rights against those of farmers. Significantly, farmers' rights are acknowledged in the very title of India's law, and a chapter of the Act is devoted to farmers' rights. Under the PPVFR Act, a farmer has the right to save, use, sow, resow, exchange, share, or sell seeds, *including from protected varieties*, as well as harvested materials, "in the same manner as he was entitled before the coming into force of this Act" (Article 39 iv, emphasis added).[15] The only restriction is that a farmer cannot sell branded seeds of a protected variety if they are labeled as such. This provision is understood as meaning that farmers can sell seeds in a generic form without a label but cannot compete with breeders and seed companies by selling under a brand name (Cohen and Ramanna 2007). Safeguarding farmers' right to sell seeds of protected varieties was the demand most fiercely resisted, and India remains one of the few countries worldwide to have such a provision (Sahai 2002).[16]

Under the PPVFR Act, farmers are recognized as breeders alongside public and private breeders and are entitled to IP protection of their varieties. Farmers' varieties are defined as those that have been traditionally cultivated and developed by farmers in their fields, and about which farmers possess common knowledge. The PPVFR Act includes several innovative provisions pertaining to farmers' rights. For example, farmers cannot be held responsible for infringing breeders' rights if they can demonstrate that they did so unknowingly, a provision meant to protect farmers who are not aware of the breeders' rights legislation. Moreover, seed companies are obligated to inform farmers of the expected yield of their varieties, and in turn farmers are entitled to compensations if the seeds do not perform as advertised. The Act also includes provisions for benefit sharing. Farmers who are engaged in the conservation and improvement of genetic resources are entitled to receive benefits through a national Gene Fund. On registering varieties, private and public breeders are obligated to declare if they have used genetic resources maintained by Indigenous or farmers' communities in the process, and these communities are entitled to receive benefits.

While farmers enjoy significant rights under the PPVFR Act, important questions remain as to the effectiveness of this approach. Almost two decades after the legislation was passed, the Act has not produced tangible results in terms of either the protection of farmers' rights over genetic resources or the preservation of agrobiodiversity (Peschard 2014; Kochupillai 2016). Farmers have not been able to avail themselves of the compensation provision in situations of crop failures. Despite high rates of registration for farmers' varieties,[17] benefit sharing is not yet a reality, and farmers who register their varieties derive neither benefit nor assistance.

In addition to enacting plant breeders' rights legislation, India also revised its legislation on industrial property to comply with the TRIPS Agreement by introducing successive amendments to its Patents Act in 1999, 2002, and 2005. The Indian Patents Act of 1970 allowed patents on processes but not on products and excluded plants and agricultural methods.[18] In other words, the process used to manufacture a plant variety or a pesticide was patentable, but not the seeds or pesticides themselves. With the amendments, both processes and products became patentable. Patents were allowed on microorganisms and on microbiological, biochemical, and biotechnological processes. This meant that genetic engineering processes as well as genetically engineered microorganisms could now be patented.

However, under pressure from civil society, India included a more elaborate definition of exclusions to patentability than most countries had established. Article 3(j) of the Patents Act stipulates that microorganisms are patentable, but explicitly excludes the patenting of "plants and animals, in whole or any part, including seeds, varieties and species, and essentially biological processes for production or propagation of plants and animals" (GoI 1970). The Act also excludes discoveries and any invention derived from traditional knowledge. These provisions were hailed as a victory by civil society. However, as Philippe Cullet (2005b) argued at the time, the significance of these exceptions was not clear. Indeed, court cases in the United States and Canada showed that even if patents are not allowed on higher life forms like plants, companies have successfully claimed de facto rights over the plants that incorporate a patented gene. If Indian courts chose to interpret a patent owner's rights as extending to any plants containing the protected microorganism, regardless of whether the plant itself was patentable—as the Canadian Supreme

Court had done in the Schmeiser case (see Introduction)—it would render these exceptions moot. In other words, the scope of these exceptions would depend on whether Indian courts interpreted the issue according to traditional patent law or instead drew a distinction between industrial goods and living organisms. Despite its shortcomings, India's unique legislation would come to play an important role in worldwide debates around intellectual property, biotech seeds, and farmers' rights.

By amending its Patents Act and enacting the PPVFR Act, India met its obligations under Article 27.3(b) of the TRIPS Agreement. What happened next is testimony to the relentless pressures exerted on countries in matter of IP rights in agriculture (Dutfield 2011). In May 2002, less than a year after the adoption of the PPVFR Act, the Indian cabinet expressed interest in joining UPOV. This, however, would have negated what had been achieved with the PPVFR Act, since UPOV did not include provisions concerning farmers' rights. The move in the direction of UPOV—in direct contradiction to everything that had been accomplished by way of national legal sovereignty in the area over the previous seven years— was probably due to strong external pressure on the government. It is plausible to assume that UPOV was keen to get a major country like India on board and prevent its sui generis legislation from becoming a model for other countries in the process of introducing plant variety protection legislation. In any case, Gene Campaign, an Indian NGO that had been involved in drafting the PPVFR Act, filed a public interest lawsuit challenging the government's decision to join UPOV on the grounds that India was under no obligation to join the organization, and that doing so would constitute a violation of its own legislation (the PPVFR Act and the Constitution) as well as of the Convention on Biological Diversity (CBD) and the Plant Treaty, of which India was a signatory (Gene Campaign 2003). In response to the public interest lawsuit, the government backtracked and denied its intention to join UPOV (Peschard 2014; Peschard and Randeria 2020).[19]

During the negotiation of the TRIPS Agreement, both Brazil and India were highly vocal in asserting their interests internationally. Yet when the time came to incorporate the TRIPS Agreement into domestic legislation, their governments did not fully exploit the room available to develop sui generis legislation adapted to their national circumstances and policy objectives.[20] The Indian government initially introduced a plant variety protection law based on UPOV 1978, and civil society itself was the driving force behind the PPVFR Act. As for Brazil, it passed legislation based on the 1978 Act of the UPOV Convention, which provides for stronger plant breeders' right than required by the TRIPS Agreement.

Although neither Brazil nor India took full advantage of the flexibilities available under the TRIPS Agreement, their legislation in matters of plant variety protection and patents still differs significantly from that of the United States. The latter arguably provides the most expansive IP protection for plants worldwide: exclusions from patentability are narrowly defined, and farmers' rights are not explicitly recognized (Sease and Hodgson 2006).[21] In Brazil and India, in contrast, the legislation includes broader exceptions to patentability and seeks to balance the rights of plant breeders against those of farmers. These differences reflect varying views of the patentability of life forms, as well as of the limits and social function of intellectual property. In addition, the deliberately vague wording of Article 27.3(b)—the key to achieving consensus during the negotiations on the controversial extension of IP rights to plant-related inventions—was carried over into domestic legislation. As global corporations such as Monsanto entered the Brazilian and Indian markets starting in the first decade of the twenty-first century, legislative differences among countries, coupled with the legal uncertainties surrounding the exact scope of Article 27.3(b), proved fertile ground for legal disputes to rage around intellectual property and biotech seeds.

2

CHALLENGING ROYALTIES ON ROUNDUP READY SOYBEAN

As [Monsanto] is amoral, it will do anything for profits.
—Luiz Fernando Benincá, Brazilian soy grower and litigant

In the early 1970s, Monsanto patented a compound called glyphosate and commercialized it as an herbicide under the brand name Roundup. Glyphosate kills plants by inhibiting an enzyme called EPSPS, responsible for manufacturing essential amino acids. By the mid-1980s, Roundup had become one of the best-selling broad-spectrum herbicides on the market—and Monsanto's main source of income (Charles 2001).

A decade later, scientists working at one of Monsanto's Roundup manufacturing plants found that some bacteria had developed the ability to resist glyphosate in the factory waste ponds (Charles 2001). By isolating genetic material from these bacteria and transferring it to the soybean through genetic engineering, scientists working for Monsanto developed a soybean plant that could withstand the application of Roundup herbicide. The genetically engineered gene that conferred tolerance to Roundup was patented as Event CP4 EPSPS and marketed as Roundup Ready (RR) soybean.[1]

At the time, like many farmers across the world, Brazilian farmers were having a difficult time controlling weeds in soybean plantations. As recalled by Luiz Fernando Benincá, a large rural producer who had been growing soybeans in Southern Brazil since 1973:

Everyone here knew Monsanto very well because Monsanto discovered the total herbicide glyphosate. . . . At that time, we were spending a fortune to control pests because we were using two or three herbicides of different brands, and sometimes we would even mix them. It cost a real fortune to control pests, it couldn't go on like that. So when Monsanto created Roundup—wow!—we were amazed. And when they developed Roundup-resistant soy through genetic modification in the United States—my God!—we all went crazy, thinking "This will save us!" (Interview #33)[2]

Salvation, however, proved short-lived. As he continues: "Nature is strong, thank God. And that's what happened: the glyphosate ended up being selective because of the massive use of this one product all the time. So, the pests that hadn't been important before became a problem, because they'd been left all alone. They developed prodigiously and so we had to use more and more products. Now, we're using a bunch of other pesticides because glyphosate no longer does the job" (Interview #33).

With such intense selection pressure, the development of weed resistance to Roundup was only a matter of time. Far from breaking away from the pesticide treadmill, the introduction of transgenic crops like RR soybean shifted the treadmill into high gear.

While farmers may have adopted RR soybean because it facilitates management on the farm, Monsanto had other motives for developing the technology. In the mid-1990s, Roundup was the company's flagship product. The looming expiration of its US patent in 2000 meant that other companies would be allowed to market generic versions, depriving Monsanto of its most lucrative source of income. From a business perspective, RR crops were a masterstroke: they created an exclusive market for Roundup herbicide when the latter was set to enter the public domain.[3]

A PRIVATE INTELLECTUAL PROPERTY REGIME: MONSANTO'S ROYALTY COLLECTION SYSTEM

In 1998, Monsanto applied to the National Technical Commission on Biosafety (CTNBio) to have RR soybean approved for production on a commercial scale. At this point, the approval process started to derail. Invoking the precautionary principle, a consumer advocacy group obtained a court injunction prohibiting the government from authorizing the sale and planting of RR soybean in the absence of environmental impact assessments and of food safety and labeling norms.[4] The injunction introduced a de facto judicial moratorium on transgenic crops.

Meanwhile, farmers in the southernmost State of Rio Grande do Sul were planting RR soybeans smuggled from Argentina and putting pressure on the Brazilian government to authorize these varieties. In March 2003—in one of his first major decisions in office—President Lula da Silva signed a presidential decree authorizing the commercialization of illegal RR soybeans, in effect lifting the judicial moratorium in place since 1998 (RFB 2003). A second and a third presidential decree came on the heels of the first. In 2005, after a seven-year legal and political battle, RR soybeans were definitively approved with the passing of a revised Biosafety Act regulating all activities involving GMOs (Peschard 2010).[5]

When the second presidential decree was passed in 2003, Monsanto moved swiftly to implement a system for the collection of royalties on RR soybean. As Felipe Filomeno (2014, 90) observes, "the decree stated that soybean growers were exclusively responsible for the cultivation of RR seeds, including obligations related to the 'occasional rights of third parties'—a hint on Monsanto's [intellectual property] rights."

In a notice to soybean growers published in major Brazilian newspapers in September 2003, Monsanto stated:

Considering that soy planting must begin within a few weeks, Monsanto advises farmers that the planting of Roundup Ready® Soybeans (transgenic soy) continues to be suspended by virtue of the decision of the TRF [Brazilian Regional Federal Court] of September 8, 2003.

Independent of the commercial release, farmers who plant Roundup Ready® Soybeans must make provisions to pay for the use of the technology at the time of selling the crop. (Monsanto 2003)

This royalty collection system was unique worldwide and came as a complete surprise to soybean farmers.[6] Indeed, the conventional practice in the seed industry had long been to include royalties in the sale price of seeds rather than to charge them on the harvested product.

CORREIO DO POVO · *GERAL* · TERÇA-FEIRA, 16 de setembro de 2003 — 5

Ministro Ruy Rosado é homenageado no TJ/RS

Destacado seu exemplo ao Judiciário nacional

O ministro Ruy Rosado de Aguiar Júnior, recentemente aposentado do Superior Tribunal de Justiça (STJ), foi homenageado por sua dedicação ao Judiciário, em sessão solene do Tribunal de Justiça do Estado (TJ/RS), no início da tarde de ontem. Além de ministro, ele foi promotor de Justiça, juiz do Tribunal de Alçada, desembargador do Tribunal de Justiça, corregedor-geral da Justiça e coordenador-geral da Justiça Federal. O retorno do ministro ao Estado foi saudado na cerimônia pelo desembargador Alfredo Guilherme Englert, que, em 1994, havia feito o discurso de despedida, quando da partida do colega para Brasília, onde assumiu o cargo no Superior Tribunal de Justiça.

Englert lembrou a atuação do homenageado na presidência da 5ª Câmara Cível do TJ/RS e o seu empenho pela implantação dos Juizados Especiais. Englert defendeu que os magistrados pertencem ao núcleo essencial do regime democrático, porque sua autonomia é sua bordinada somente à Constituição, o que considera um obstáculo tanto ao liberalismo exacerbado quanto às transformações revolucionárias de qualquer governo. O discurso foi concluído com elogios ao trabalho do ministro, considerado marco referencial de julgador em todo o país. "O trabalho de vossa excelência extrapolou o prédio do STJ para se ramificar por todo o Brasil", ressaltou.

CRISTIANO SANT'ANNA

Reencontro com os antigos conhecidos

Para o homenageado, a cerimônia foi um momento de alegria e de reencontro com antigos conhecidos. Ruy Rosado observou que sempre soube que retornaria ao TJ/RS para prestar contas da sua jurisdição. "Procurei não me afastar da tradição da magistratura gaúcha, cujo conceito e significativo prestígio comprovei a todo instante, em qualquer auditório do país, e tentei ser nada mais do que isso: um juiz do Rio Grande", declarou. "O que fiz de acerto foi apenas uma tentativa de expressar o que um juiz deste tribunal faria no mesmo lugar".

Na sua avaliação, a Justiça atravessa um momento difícil, pela incerteza das reformas do Judiciário e da Previdência, agravadas pela incapacidade de resposta útil à crescente demanda. Entretanto, o ministro considera que a Justiça gaúcha busca qualificação e aprimoramento, citando a criação dos Juizados Especiais, que considera a melhor inovação para a prestação da jurisdição feita no país. "Temos, enfim, um tribunal que administra recursos com competência, a ponto de dotar todas as comarcas de excelentes condições físicas. Por tudo isso, um Estado que se destacou pelo pioneirismo na busca de tantas soluções inovadoras certamente encontrará meio e modo de superar as dificuldades presentes", concluiu.

Fausto apóia redução da jornada

O presidente do Tribunal Superior do Trabalho (TST), ministro Francisco Fausto, afirmou ontem que apoia integralmente a proposta de redução da jornada semanal, de 44 horas para 40 horas, em discussão no Fórum Nacional do Trabalho. "Sou absolutamente favorável à redução da jornada e avalio que ela implicará a ampliação do mercado de trabalho no país", afirmou.

Segundo ele, a medida deve ser adotada sem redução salarial. "Já se registra hoje uma queda de 8% da massa salarial do país, o que é alarmante, e reduzir ainda mais os salários seria inadmissível." Para Fausto, a compensação às empresas, por conta da redução de horas trabalhadas, deveria ser feita pela via fiscal. O ministro lembrou que, na França, a redução da jornada de trabalho causou polêmica, pois não gerou o correspondente aumento do nível de emprego. "No Brasil, não creio que isso vá ocorrer", observou. "Teremos condições de gerar novos empregos com a redução da jornada e, sobretudo, com algumas medidas paralelas, como a proibição de horas extras exageradas. Até porque, se se reduz a jornada de trabalho e se paga horas extras aos empregados, isso pode destruir a intenção do governo de ampliar o emprego."

Para o presidente do TST, a preocupação do setor empresarial, que é mudança geral alta de preços, pode ser contornada com um sistema de compensação tributária.

Parcerias dão trabalho a presos

O Tribunal de Justiça (TJ/RS) contará com o apoio de mais 27 entidades no projeto "Trabalho para a vida", da Corregedoria-Geral da Justiça. As novas parcerias, que serão firmadas hoje, se somarão às três entidades que estão renovando a participação nessa iniciativa. O presidente do Tribunal de Justiça, desembargador José Eugênio Tedesco, e o secretário da Justiça e da Segurança, José Otávio Germano, participarão da solenidade, às 13h30min, no Tribunal de Justiça.

A união tem como objetivo gerar vagas de trabalho e proporcionar ensino profissionalizante aos apenados. O termo de compromisso a ser assinado pelas entidades públicas e privadas prevê a criação de 20 cooperativas de trabalho ou produção destinadas aos detentos do sistema prisional gaúcho.

2.1 Monsanto notice to Brazilian soybean producers, September 2003

2.2 Monsanto newspaper ad, February 2005

With the imminent passing of the revised Biosafety Bill, Monsanto published another full-page ad in the popular newspaper *Correio do Povo* in February 2005, informing farmers of its new policy:

Monsanto invested heavily in tests and research to develop transgenic soy. And it continues to invest 500 million dollars a year to bring in new technologies for the future such as insect-resistant soybeans, drought-tolerant soybeans, and soybeans that is even healthier. For this reason, if you opted for the advantages of planting transgenic soy for the 2004–2005 crop, you must know that royalties for the use of the Roundup Ready technology will be charged *at the time of selling your crop*. [emphasis added] (Monsanto 2005)

The royalty collection system implemented by Monsanto in Brazil amounts to a *private* intellectual property (IP) regime, in the sense that it bypasses public policy-making and regulations (Filomeno 2014). Monsanto's royalty collection system has been denounced by litigants as a form of self-regulation through which transnational corporations set standards and norms for their own transactions and take the necessary steps to implement them (Filomeno 2014). As Monsanto's business manager admitted frankly at the time: "By virtue of the [Plant Variety Protection Act], which guarantees to farmers the right to save seeds, I believe we will be operating two systems in Brazil: one in which (royalties) are collected after the harvest, in the case of illegal soy, and the other in which (royalties) are collected when the seeds are sold, in the case of legalized planting" (Kassai 2005). This statement is a rare admission on the part of Monsanto that the Brazilian Plant Variety Protection Act (PVP Act) does in fact guarantee the right to save seeds for replanting. At the same time as the business manager recognizes this right, he refers to these seeds as "illegal" and announces a charge that will infringe on this right. Monsanto refers to royalties collected on grain, as opposed to seeds, as an "indemnity for the unauthorized use of a patented technology" (ClicRBS 2005).

The cornerstone of Monsanto's royalty collection strategy was a General Agreement on the Licensing of Intellectual Property Rights on Roundup Ready® Technology (Monsanto n.d.), an agreement that Monsanto required Brazilian seed producers to sign so that they could produce and distribute RR soybean. By signing this agreement, seed producers entered into a contractual relationship with Monsanto and became ultimately responsible for collecting royalties from farmers. The agreement stipulated that the seed producer "agrees to act on behalf of Monsanto as its representative vis-à-vis soybean farmers, in compliance with the applicable provisions of the Civil Code, to implement the licensing of RR technology found in RR seeds used by soybean farmers in their fields; this includes a commitment, on the part of the seed producer, to collect royalties from soybean farmers and pay them to Monsanto according to the terms of this Agreement" (Monsanto n.d.).

The royalty collection system that Monsanto implemented in Brazil involved devising a computerized system to keep track of farmers' seed purchases and grain sales. In fact, much of the 25-page agreement consists of a detailed description of the conditions of implementation. The preamble of the agreement states that "Monsanto developed the technology and claims to hold, in Brazil, intellectual property rights on the gene sequence that confers soybean resistance to glyphosate-based herbicides (otherwise known as 'RR Technology')" (Monsanto n.d.). It stipulates that Monsanto's rights extend not only to "the production and commercialization of RR seeds" but also to "the planting and commercialization of RR soybean" (Monsanto n.d.). And thus, because "it is common among soybean farmers to save seeds for sowing or planting," Monsanto intends to collect royalties at the time of the acquisition of seeds as well as for the authorization to use reserved seeds (Monsanto n.d.). Reserved seeds (*sementes reservadas*) is the expression used by Monsanto to refer to seeds saved by soybean growers for their own use, that is, for replanting (Monsanto n.d.). The company defines "royalties" as its "remuneration value for RR seeds acquired by soybean farmers and/or for the use of reserved seeds by soybean farmers" (Monsanto n.d.).

The Brazilian Seed Producers Association (ABRASEM) initially warned its members that "all seed producers who received or will receive a copy of the Monsanto agreement should consult their lawyers before signing it . . . because . . . many elements are legally incomprehensible" (Reis 2005). The agreement did not specify which patent(s) covered the technology in Brazil. The claim that Monsanto's IP rights extend to commercialization and production was also legally questionable, given Brazil's legislation. Indeed, under the PVP Act, plant breeders' IP rights apply to seeds but do not extend to harvested materials.[7] Following further negotiations with Monsanto, ABRASEM gave in and accepted the agreement, in exchange for a slightly greater share of the royalties (Reis 2005). A number of ABRASEM members denounced the agreement, among them the Seed Association of Rio Grande do Sul (APASSUL).

By 2006, Monsanto had successfully implemented a private royalty collection system that ensured that it collected royalties, no matter the origin of the seeds. This "dual remuneration scheme" is a shrewd system: the farmer who does not pay royalties when purchasing the seeds is nevertheless forced to pay royalties when they sell their harvest. In practice, this system eliminated farmers' right to save seeds.

Around the time I conducted research for this book, a soybean farmer who showed up at the Bianchini processing plant in Rio Grande do Sul to sell his harvest of Intacta soybeans (the second-generation genetically modified soybean varieties launched in 2013) was met by a poster that read:

Attention, Mister Supplier!
By virtue of industrial and intellectual property law, it is compulsory to conduct tests to identify these varieties.
For Intacta Soy to be accepted, the supplier must:
1. Possess credits from the purchase of certified seeds; or
2. Agree to a 7.5 percent discount on the value of the shipment of this variety*;
3. Agree to a 7.5 percent discount in kind on the net weight of the shipment in the event of a subsequent transfer to third parties.

Do not waste time. Inform the recipient if the soy is Intacta. Do not mix Intacta soybeans with other varieties.

*Value of the shipment = the market price at the time of the sale of the soybeans. (Bianchini n.d.)

In Canada and in the United States, the legal keystone of Monsanto's IP strategy is a Monsanto Technology Stewardship Agreement signed by farmers when they buy seeds from the seed dealer. This agreement has served as the legal basis for hundreds of patent infringement lawsuits that Monsanto has filed against farmers (Center for Food Safety and Save Our Seeds 2013). However, as Monsanto came to acknowledge when it entered the Brazilian market in the early 2000s, this scheme was simply not practicable in Brazil. In a context where seed saving was far more widespread than in the United States, Monsanto simply lacked the means to investigate or prosecute each and every farmer. Moreover, such practices would certainly have backfired and unleashed retaliation against the company. Monsanto's solution was to devise a system whereby it delegated the responsibility for collecting royalties to intermediaries in the production chain—seed producers and grain elevators. This system depended on their collaboration, which Monsanto was able to secure.

The company succeeded in implementing this unique royalty collection system for a number of reasons. In the early years, RR soybean was touted internationally as a revolutionary technology and even a panacea. In this context, there was a real fear among soybean growers—like

2.3 Poster on royalties at soybean processing plant, Brazil

the farmer quoted at the beginning of the chapter—of being left out, and this sentiment was ably exploited by Monsanto. The company also took advantage of the lack of expertise in Brazil in the area of intellectual property for plant varieties and biotechnological traits. Patents on agricultural biotechnology were becoming a new reality everywhere, but in countries like Brazil, plant variety protection was also a new reality. Finally, Monsanto was able to gain the support of the main agricultural federations, without whose collaboration it could not have put the royalty collection system in place. Many soybean farmers were left with a bitter feeling after their own leadership chose to side with Monsanto rather than defend their interests. Their sense of betrayal sowed the seeds, so to speak, of future litigation.

THE PASSO FUNDO CLASS ACTION

In January 2005, Cotricampo—a rural cooperative in the northwest region of Rio Grande do Sul, representing 8,000 soybean farmers— obtained a preliminary injunction suspending the collection of royalties on harvested soybean. In the first ever decision on the matter in Brazil, the judge ruled that the patent on Monsanto's RR technology covered the seeds but did not extend to production, and thus suspended the royalty charged on every bag of grains (1.20 BRL or 0.49 USD, at the time). As he wrote in his decision, "the intellectual (property) rights, including those related to genetic modification, granted under the PVP Act, only extend to the plant reproductive material and obviously not to the entire soybean production" (Consultor Jurídico 2005). The injunction that Cotricampo obtained prompted other soybean farmers to file individual and collective lawsuits on similar grounds.

However, a month later, another judge revoked the injunction, suspending the right to save seeds when it came to transgenic varieties:

Article 10 of [the PVP Act], which regulates intellectual property in plant varieties specifically, does not apply; indeed, even if we deem that this Act voided the rights guaranteed by the [Industrial Property Act], which is quite arguable, this could only apply if the farmer had paid royalties at the time he first acquired the seeds; this is obviously not the case, since it is public knowledge that all transgenic soybean seeds entered the country illegally, and were not sold (here) by the Respondent, who, for this reason, did not charge royalties. (*Cotricampo v. Monsanto*, 2005)

In his decision, the judge eschewed the vexed issue of determining whether either the Industrial Property Act or the PVP Act applied, and simply ruled that farmers' right to save seeds was void because they had not initially purchased commercial seeds. In reasoning reminiscent of the decision of the Supreme Court of Canada in *Monsanto v. Schmeiser*, the judge ruled that the seeds belonged to Monsanto, no matter how they had landed in farmers' fields and independently of the fact that it was illegal to plant them at the time. In yet another lawsuit around the same period, the judge shifted the burden of proof, ruling that "the Defendant [Monsanto] would only be authorized to charge royalties if it could prove that it sold the seeds, a requirement that it cannot fulfill, but which cannot be attributed to the farmer" (Costa 2005).

As these examples show, early challenges to the royalty collection system produced contradictory interpretations. Local and higher-court judges for the first time had to confront complex legal issues that required an understanding of plant breeding, genetic engineering, biotech patents, and plant breeders' rights. They were in uncharted waters, as there was yet no legal precedent on these issues in Brazil. To complicate matters, no one knew the status of Monsanto's Brazilian patents.

These court cases caught the attention of Luiz Fernando Benincá, the soybean grower quoted earlier. Benincá lived in the region of Passo Fundo, a regional hub at the heart of the soybean-growing northern region of the State, and felt deeply dissatisfied with the fact that Monsanto charged royalties on harvested materials. Emboldened by the fact that Cotricampo had obtained an injunction, he filed his own individual lawsuit in June 2005. Like the cooperative, he obtained a preliminary injunction in his favor.

At the time, Benincá was president of the Passo Fundo rural union, and he decided to take the issue to the Agricultural Federation of the State of Rio Grande do Sul (FARSUL), which represents large rural producers and employers. With the injunction in hand and accompanied by his lawyer, he went to a regional FARSUL meeting in the city of Não-Me-Toque in July

2005. There, he took the microphone to explain the issue and suggest that FARSUL take action. To his surprise, the federation's president responded by saying that royalties ought to be paid, otherwise Monsanto would retaliate by detaining soybean shipments at sea. As Benincá recounts, "Stop a ship on the open sea? That's a ridiculous idea, you know. . . . At that point, I knew that my suspicion was founded and that FARSUL was complicit in this." When he insisted on speaking, the FARSUL president cut him off. As the farmer says bitterly, "No one could have imagined that our own federation would lend itself to organizing the collection of royalties on behalf of Monsanto" (Interview #33).

Benincá eventually lost his individual lawsuit.[8] Despite this setback, he did not give up. He remained convinced that the royalty collection system rested on shaky legal ground. Moreover, he felt that he had a more solid case because he now had a better grasp of the legal issues than when he first set out to challenge the royalty collection system four years earlier. Knowing full well that he could not count on the support of his federation, he decided to approach his local rural union in Passo Fundo.

On April 9, 2009, the rural union of Passo Fundo filed a class action lawsuit against Monsanto Technology LLC and Monsanto of Brazil in a civil court in Porto Alegre, the State capital. In their petition, the plaintiffs argued that the royalty collection system was arbitrary, illegal, and abusive, and that it hurt the collective rights of millions of farmers.

The plaintiffs asked the judge to take into consideration the fact that food production was a matter of public interest (Sindicato Rural de Passo Fundo-RS 2009). More specifically, they argued that the dual remuneration system put in place by Monsanto contravened Brazilian legislation. By enacting the PVP Act, they argued, Brazil had explicitly rejected the protection of plants under patent law. Moreover, by adhering to UPOV 1978 as opposed to UPOV 1991, Brazil had opted *not* to extend IP rights to harvested materials. The plaintiffs asked the civil court to reaffirm the right of farmers, under the PVP Act, to save seeds from their crops for replanting on their farms; to sell their harvest as food or raw material without paying royalties; and, in the case of small farmers, to give away or exchange seeds among themselves.

The rural union claimed that the amounts collected as royalties for RR soybean technology were abusive and represented the unjustified enrichment of a private party at the expense of farmers. According to the court petition, an estimated 140 million BRL (58 million USD) was collected in royalties on RR soybeans in the State of Rio Grande do Sul that year, and 1 billion BRL (500 million USD) was collected in the whole of Brazil (Sindicato Rural de Passo Fundo-RS 2009).[9] The rural union rejected Monsanto's argument that farmers had agreed of their own free will to the royalty collection system through the agreement reached with FARSUL. Rural unions were farmers' only legitimate representatives, yet they had not been consulted on the royalty collection system, let alone been allowed to vote in assembly.

The rural union therefore asked the judge to declare the royalty collection system implemented by Monsanto illegal and to set the percentage of royalties in line with royalties on conventional plant varieties. It also demanded that Monsanto pay back the amounts unduly collected on harvested grains since the 2003–2004 harvest. Finally, the rural union asked the judge to establish the value of the class action at 1 billion BRL (500 million USD) and to order the judicial deposit of the amounts in dispute in an escrow account awaiting a final ruling.

This legal action was started by one rural union, Passo Fundo, but the idea quickly picked up momentum. Other rural unions—including Santiago and Sertão—joined the action shortly after.[10] These unions were dissident chapters that were not in step with the federation leadership. Significantly, FETAG-RS—a state federation of 350 family-farming local unions—voted unanimously to join the action in June 2009 (FETAG 2009). While the rural unions represent large farmers and rural employers, FETAG-RS represents small farmers and rural workers. In the polarized Brazilian agrarian landscape, this changed the profile of the case: no longer simply a dispute about profits among powerful economic actors, it came to encompass the rights and livelihood of small farmers (Interview #38). In addition, a number of other groups also joined the action as interested parties, including farmers' unions and associations, public institutions such as the Public Prosecutor's Office of Rio Grande do Sul[11] and the Brazilian Patent Office (INPI), seed producers, and agbiotech industry organizations.

In April 2012, Judge Giovanni Conti from the State Civil Court delivered his decision in the class action.[12] Judge Conti ruled in favor of the rural unions, accepting their line of argument in its entirety (*Sindicato rural de Passo Fundo v. Monsanto*, 2012). In his decision, he reaffirmed the right of all farmers, small as well as large, under the PVP Act, to save seeds for replanting without paying royalties; and the right of small farmers to exchange or give away seeds among themselves. The judge determined that Monsanto's IP rights had been exhausted by licensing its technology to seed producers and selling seeds to farmers and, therefore, Monsanto was not entitled to collect royalties on harvest. Consequently, he suspended the collection of royalties on harvested grain with immediate effect, subject to a daily penalty of 1 million BRL (513,000 USD). The judge also ruled that Monsanto had to pay back the royalties it had collected on harvested grain since 2003–2004. Monsanto immediately appealed the decision.

In a parallel development, Monsanto and the rural unions each filed a special appeal before the Superior Court of Justice (STJ)[13] to challenge specific dimensions of a preliminary injunction by the Court of Justice of Rio Grande do Sul. For Monsanto, it was an attempt to terminate the case. For the rural unions, however, it was an opportunity to expand its scope. The STJ delivered its decision on June 12, 2012, ruling against Monsanto on both counts.

Monsanto's appeal challenged the legitimacy of rural unions to file a class action on behalf of farmers, arguing that the matter had to be looked at as an individual relationship between each soybean farmer and the company (COAD 2012). The Court denied this request and confirmed the admissibility of the action, stating that "the present action was not undertaken solely to defend the labor interests of the members of the association. The action was filed with the objective of protecting, in a broad manner, the rights of all farmers who work with transgenic RR soybeans. In other words, the action was filed in the interest of the professional category as a whole" (*Monsanto v. Sindicato Rural de Passo Fundo*, 2012). The Court added that the discussion around royalties was socially relevant because it was reflected in food prices.

The union's appeal concerned the scope of the case. At Monsanto's request, the Court of Justice of Rio Grande do Sul had reduced the scope of the decision to the State of Rio Grande do Sul, the Court's territorial jurisdiction. In response, the rural unions asked the STJ to reestablish the national relevance of the case. The STJ accepted the plaintiffs' argument that the royalty collection system affected all Brazilian soybean growers equally and that any decision should therefore be national in scope (COAD 2012).

However, the STJ granted Monsanto's objection to both the suspension and judicial deposit of royalties. This meant that farmers would continue paying royalties until the STJ delivered a final decision in the case.[14] As a consequence of the STJ ruling, the case was allowed to proceed, and its scope was deemed to apply to approximately 4,000,000 farms (COAD 2012). The amount of royalties in dispute was estimated at 15 billion BRL (7.7 billion USD) (COAD 2012).

Meanwhile, the class action was proceeding. In a two-to-one decision delivered in September 2014, the Court of Justice of Rio Grande do Sul overturned the lower court decision by Judge Conti (*Monsanto v. Sindicato rural de Passo Fundo*, 2014). The Court offered an interpretation narrowly grounded in patent law. It ruled that as a product of genetic engineering, RR soybean came under the exclusive protection of the Industrial Property Act, and those who opted to use RR soybeans had an obligation to compensate the patent holder for the use of the technology.

The rural unions then lodged a special appeal before the STJ. The Court determined that the central question in the case was whether farmers could avail themselves of the right to save seeds from their crops for replanting, in the case of patented soybean. In October 2019, the Court's nine judges unanimously ruled in favor of Monsanto (*Sindicato Rural de Passo Fundo v. Monsanto*, 2019). Judge Buzzi opened with a lengthy statement on the importance of soybean in Brazil's agribusiness exports, revealing the extent to which economic considerations had weighed in on the decision. The Court argued that the exhaustion principle does not apply in cases in which a patented product is used for multiplication and commercial propagation. The Court also determined that the exceptions to plant breeders' rights established in Article 10 of the PVP Act only apply to the holders of a Plant Variety Certificate (that is, to plant breeders). These exceptions, the Court stated, were not enforceable against the holders of patents on processes or products related to genetic engineering when the object of the patent is found in plant reproductive material. This ruling significantly restricted farmers' rights to save seeds from genetically engineered plant varieties throughout Brazil.

MONSANTO'S INTELLECTUAL PROPERTY RIGHTS TO ROUNDUP READY SOYBEAN

In the early years of the royalty collection system, no one knew which patents protected RR soybeans in Brazil—since the RR soybean was essentially a "black box" in terms of patent information (Rodrigues, Lage, and Vasconcellos 2011). Monsanto did nothing to clarify the situation (Souza Junior 2012). The company made vague statements about RR soybeans being protected by a range of IP rights, without ever specifying patent numbers, even in its licensing agreements. A lawyer for the Federation of Agriculture and Livestock of Mato Grosso (FAMATO) who analyzed a licensing agreement for Intacta soybeans pointed out that Monsanto "never disclosed the patent number, in spite of many requests, including from the courts" (FAMATO 2013).[15] Another corporate strategy consisted of muddying the water by providing extraneous information. As one legal expert observed with regard to the Passo Fundo class action: "Patent PI-9708457-3 has nothing—even remotely—to do with the subject of Class Action 82–2012, and must only have been introduced to the procedure to obscure the central legal matter of the lawsuit" (Barbosa 2014, 355).

Prompted by the legal actions, a number of lawyers and legal researchers started to delve into the subject. The expert advice requested by Judge Conti in the Passo Fundo class action was one of the early inquiries that shed light on the status of Monsanto's Brazilian patents. As part of his investigation, the judge asked Luiz Carlos Federizzi, a plant breeder from a public university, for an expert opinion on Monsanto's patent rights over RR soybean technology in Brazil. Federizzi obtained Monsanto's application file for commercial authorization from the regulatory agency, which included information about its patents. Since Monsanto had applied for a patent in Brazil through the pipeline mechanism (see chapter 1), the expert then looked at the corresponding US patent application.[16]

The expert made a number of findings. First, he concluded that of the five patents submitted by Monsanto to the court, only one was relevant to the case: "After the detailed analysis of the patents concerned in this trial, along with consultations with the [National Institute of Industrial Property] and the US Patent and Trademark Office, it is clear that the patent corresponding to the one in the Monsanto application approved by the [Brazilian regulatory agency] CNTBio is PI 1100008–2. Therefore, the other patents filed in the present case are either included in or were superseded by patent PI 1100008–2 and do not need to be considered here" (Federizzi 2011, 16).

Moreover, the expert confirmed that this patent had expired in Brazil on August 31, 2010—that is, twenty years after the patent application was filed in the United States. The expert also noticed that the claims contained in the two patent applications differed: some claims in the US patent were modified or simply removed from the Brazilian pipeline patent. The reason is plain to see: while a gene can be patented in the United States, patenting a gene is not allowed under the Brazilian legislation. Consequently, Monsanto removed all items related to genes and instead focused on the process for controlling the weeds around a soybean plant (Interview #39). The number of claims increased significantly: while the original US patent contains eight claims, the corresponding Brazilian patent contains no fewer than 73 claims (Federizzi 2011). This raises questions about the patent's validity, since one of the conditions for obtaining a patent through the pipeline mechanism is that the claims must be the same as those in the original patent. The claims are the most important part of a patent, because they define the nature (process or product) and the scope of protection granted to an invention.

In 2013, FAMATO and the Soybean Producers Association (APROSOJA) commissioned another legal study of Monsanto's IP rights. The resulting report by Denis Borges Barbosa, an internationally renowned Brazilian jurist specializing in biotechnological patents, was the most in-depth legal analysis so far of the patents and contractual model used by Monsanto in Brazil (Barbosa 2014).

Barbosa found that Monsanto had filed 14 patent applications with the Brazilian Patent Office related to RR soybeans and Bt cotton.[17] None of them, however, was in force when he published his report in February 2013. In some cases, Monsanto itself had withdrawn the patent application. In other cases, the Brazilian Patent Office had either dismissed or archived the patent application. In yet other cases, the patent had been granted but the period of protection had already expired. The last of the patents in force, PI 1100008–2, had expired in August 2010. Barbosa therefore confirmed Federizzi's finding that the technology had entered the public domain in September 2010 (Barbosa 2014).

Monsanto argued that it had continued to charge royalties after the patent's expiration because the extension of this patent was before the Brazilian courts. Indeed, the company had applied to the Brazilian Patent Office for an extension of the term of protection of seven of its 14 pipeline patents. The Patent Office, however, consistently denied these requests.[18] As a result, Monsanto filed multiple legal challenges against the Patent Office.

Patent PI 1100008–2 on RR soybean was no exception. Monsanto applied to the Brazilian Patent Office for a patent on tolerance to glyphosate in 1996, under the pipeline mechanism of the Industrial Property Act. The Patent Office granted patent PI 1100008–2 in 2007 (Barry et al. 2007). Since the original patent application was filed in the United States on August 31, 1990, the 20-year protection period expired on August 31, 2010.

In the United States, however, a continuation-in-part patent application had extended this patent for another four years. Continuation-in-part is a concept particular to US patent law that allows an applicant to claim enhancements to an invention already patented. The addition of enhancements thus enables the patent holder to extend the term of the original patent. This mechanism has been criticized as a form of "evergreening" that allows companies to extend their exclusive patent rights beyond 20 years. In the case of the RR soybean, for example, the first patent application in the United States was filed on August 31, 1990, and granted as US Patent RE39247 (Barry et al. 2006). Monsanto subsequently withdrew the application, filed a continuation-in-part application, and was granted another patent valid until May 2014.[19]

Monsanto wanted to benefit from the same extension of its US patent on RR soybean in Brazil. The corporation argued that the term of protection of a pipeline patent should correspond to the term of protection of the corresponding foreign patent. In 2008, it applied to the Brazilian

Patent Office for an extension of the term of patent protection until May 2014, in line with its US patent (Barbosa 2014). The Patent Office denied the extension in April 2011 (*Sindicato Rural de Passo Fundo v. Monsanto*, 2012). Monsanto then challenged the decision of the Patent Office before the Court of Justice of Rio de Janeiro.

The rules governing the term of protection for pipeline patents are clearly laid out in Articles 230 and 231 of Brazil's Industrial Property Act. Monsanto lost in the first and second instances and appealed to the Superior Court of Justice. Litigation involving the prorogation of Monsanto's pipeline patents in Brazil was disposed of by the Superior Court of Justice in June 2013. Judge Villas Bôas Cueva ruled that "The protection for foreign patents—so-called pipeline patents—is in force for the time remaining in the protection term in the country where the initial patent application was submitted, up to the maximum protection period allowed in Brazil—twenty years—starting from the date of the initial submission outside the country, even if the application is subsequently abandoned" (*Monsanto v. INPI*, 2013).

By applying to the Brazilian Patent Office for the extension of its pipeline patents and then filing multiple lawsuits against the Patent Office for denying them, Monsanto gained time and perpetuated confusion over whether its soybean varieties were still under patent protection. To a certain extent, its strategy succeeded: it collected royalties on RR soybeans until February 2013—that is, for two and a half years after the expiration of Patent PI 1100008-2. By then, Monsanto was ready to bring to market Intacta RR2 PRO, its second-generation genetically engineered soybeans. Like most second-generation GMOs, this variety is "stacked," meaning that in addition to the herbicide-tolerance trait present in RR1, it also contains an insecticidal gene known as a Bt gene. The royalties charged on Intacta were significantly higher than for RR1: 7.5 percent compared with 2 percent.

At the time Intacta was introduced, Monsanto had been granted a single patent related to this variety in Brazil, PI 0016460-7 (Fincher 2012). The Brazilian Patent Office had initially delivered a negative opinion on the patent application, raising objections based on exclusions to patentability under Article 18(3) of the Industrial Property Act. In response, Monsanto withdrew all claims that conflicted with that article and simply resubmitted the application. The Patent Office then delivered another negative opinion in which it questioned whether the invention involved an inventive step. After further explanations, the Patent Office granted the patent in October 2012 (the patent was set to expire in October 2022). The patent comprised 10 claims to a DNA sequence, to DNA constructs, to a method to express a DNA sequence, and to a method to control weeds.

Monsanto's Brazilian marketing manager for soybean was quick to emphasize that the 2022 expiry date did not reflect the "real" term of protection: "The time frame does not reflect the ultimate validity of the patent, given the fact that new applications are currently being analyzed jointly with the [Brazilian Patent Office] regarding the technology" (Folha do Cerrado 2014). Indeed, Monsanto had at least another nine patent applications related to various aspects of the technology under examination by the Patent Office. This reflects the company's practice of filing multiple patent applications in order to shore up its patent protection. In a comment on the possible revocation of patent PI 0016460–7, one legal analyst notes: "The decision would affect only one patent, and it is not clear how many of Monsanto's supplementary patents protect Intacta RR2 PRO soybean seeds" (Jurrens 2018).

When the first Passo Fundo class action was filed in 2009, little was known about Monsanto's Brazilian patents. The picture that emerged from the lawsuit was problematic. However, in their special appeal to the Superior Court of Justice, the rural unions and FETAG decided not to include issues surrounding patents because they believed this might compromise the admissibility of the case (Interview #29B). The Court could have argued, for instance, that this aspect was not included in the initial petition and that patents fell under the jurisdiction of specialized courts.

In 2017, APROSOJA-MT filed a lawsuit in a federal court challenging the validity of Monsanto's patent PI 0016460–7 on Intacta soybeans. This was the first legal challenge dealing specifically with patents, as opposed to the royalty collection system more generally. APROSOJA-MT asked the court to revoke the patent on two grounds. First, it argued that Monsanto combined already existing technology and that Intacta therefore failed to meet the innovative step criteria.[20] Second, it argued that the invention was not described in a way that would allow a skilled person to reproduce the invention once in the public domain, which is another patentability requirement. Interestingly, the Brazilian Patent Office made a submission to the court in support of revoking the patent it had granted in 2012 (Tosi 2018). It must be noted that APROSOJA-MT was seeking to revoke the patent on technical grounds, not on the grounds that such patents failed to comply with the Brazilian legislation on exclusions to patentability.

The Passo Fundo class action has had manifold repercussions. First, it spurred the development of IP expertise among soybean growers, farmer union representatives, lawyers, public servants, and even judges—something that had been sorely lacking 10 years earlier. Second, the class action shed light on the status of Monsanto's patents and, in some cases, raised troubling questions concerning their validity. Third, the class action exposed how the private IP mechanisms that the corporation had implemented for the collection of royalties on its soybean varieties subverted the domestic legislation on the protection of plant varieties. Fourth, the class action revealed how public-private partnerships often blur the line between public research and commercial exploitation, and between public and private interests. In the mid-1990s, the Brazilian Public Agricultural Research Corporation (EMBRAPA) entered into technical cooperation agreements with Monsanto for the introduction of the Roundup Ready trait into EMBRAPA's soybean varieties. It thus became a direct beneficiary of the biotech royalty system implemented by Monsanto and therefore acquired a financial stake in it. In response to the Passo Fundo class action challenging the royalty collection system, an EMBRAPA researcher said: "Although EMBRAPA has other financial sources, if the collection of royalties is interrupted then 5 to 10 million USD will be cut from our budget, which would stop some research projects" (Massarini 2012). Finally, the class action forced the judiciary to examine the conflict between patent law and plant variety protection when it comes to issues of transgenic varieties and farmers' rights to save seeds.

Around the same period, halfway around the world, litigation involving Bt cotton in India brought a similar set of issues into the limelight.

3

BT COTTON: THE PATENT THAT NEVER WAS

Fourteen years after US multinational Monsanto brought the genetically modified (GM) Bt cotton (Bollgard) to India, there is no clarity on the discovery having ever been patented in the country.
—*The Times of India* (Arya and Shrivastav 2015)

Along with soybean, cotton represented another large potential market for agricultural biotechnology companies. Monsanto developed genetically engineered cotton by inserting a gene from the soil bacterium *Bacillus thuringensis* (hence the name Bt) into the cotton genome. The Bt gene enables plant cells to produce a protein, Cry1Ac, that is toxic to major cotton pests such as the American, spotted, and pink bollworms. This genetically engineered trait was identified as Event 531 and first commercialized in the United States in 1996 under the trade name Bollgard-I, or BG-I. While Roundup Ready (RR) crops such as RR soybean are tolerant to herbicides, Bt crops such as Bt cotton are resistant to insects. These two traits, which are increasingly stacked in the same plant, are found in the totality of GM crops cultivated worldwide. In 2019, stacked traits represented 45 percent of the global biotech crop area, herbicide tolerant crops 43 percent, and insect resistant crops 12 percent (ISAAA 2019).

In India, the Genetic Engineering Approval Committee (GEAC) authorized the commercial cultivation of Bt cotton in 2002. In 2006, Monsanto introduced Bollgard-II (BG-II), which consists of two stacked Bt genes, Cry1Ac and Cry2Ab (Event 15985) and was marketed as "enhanced Bt."

From the illegal spread of unauthorized varieties to farmer suicides and the development of pest resistance, Bt cotton has been mired in controversy in India from the beginning.[1] Notwithstanding the controversy, Bt cotton spread rapidly. By 2016, it was cultivated by an estimated 7 million farmers on 10 million hectares, or 90 percent of the total cotton-producing region of the country (Bera and Sen 2016; Das 2016). In 2021, Bt cotton remained the sole genetically engineered crop authorized in India. Indeed, a public interest lawsuit brought to the Supreme Court in 2005 by Aruna Rodrigues to challenge the release of genetically modified organisms (GMOs) by the government of India in the absence of a proper biosafety protocol had prevented the approval of other GM crop varieties (*Aruna Rodrigues v. Union of India*, 2005).

A PUBLIC–PRIVATE INTELLECTUAL PROPERTY REGIME: MONSANTO'S SUBLICENSING MODEL

In the early days of genetic engineering in the 1990s, Monsanto licensed its genes to seed companies in exchange for a lump-sum payment, in effect ceding control over them. For example, in 1992, Monsanto gave the US-based seed company Pioneer the right to use the Roundup resistance genes in its soybean varieties forever in exchange for a one-time payment of half a million dollars (Charles 2001).

Monsanto initially pursued a similar strategy in India. In 2002, it offered to sell the Bt technology to the Indian government for a lump-sum payment so the government could introduce the trait into public sector cotton varieties. However, the government deemed the price Monsanto was asking for—4 crore INR, or 1,275,000 USD—too high and the deal fell through. According to a scientist from the Indian Council of Agricultural Research (ICAR): "Monsanto was ready to sell the entire technology at one go. This could have enabled free use of the technology by public sector units for just Rs4 crore. And even if the government would have further sold the technology to [the] private sector it still would have been a much cheaper deal" (Arya and Shrivastav 2015). With the benefits of hindsight, this price would indeed have been well below the amount

Monsanto has earned in royalties or "trait fees"[2] for Bt cotton in India since 2002. Monsanto does not make public how much it collects in royalties, but sources in the Indian seed industry claim to have paid over 780 million USD (5,000 crore or 50 billion INR) between 2002 and 2015 (*Economic Times* 2015).

Indeed, Monsanto eventually came up with a more profitable scheme: having farmers pay a technology fee to Monsanto, "in effect buying the new genes in a separate transaction from the seed purchase" (Charles 2001, 152). In this way, the company was licensing its genes to each and every farmer. In the United States and Canada, this was done directly by having farmers sign a Technology Use Agreement on the purchase of seeds. In Brazil and India, this was done indirectly—through sublicensing agreements signed between Monsanto and grain traders in the case of Brazil, and between Monsanto and seed companies in the case of India. These legal arrangements enabled the company to retain control over pricing but also to enforce a ban on seed saving. Importantly, these arrangements ensured that Monsanto would retain control of the technology independently of its ability to obtain or enforce patent rights. As Ian Scoones had already observed in 2006, "It is through the charging of technology license fees to partners and systematic market penetration via other players that good returns on investments in both R&D and regulatory clearance can be made, *even in the absence of strong or enforceable intellectual property protection*" (2006, 165, emphasis added).

In India as in Brazil, Monsanto faced a much more complex and fragmented rural reality than in the United States, with millions of farmers on small land holdings and intractable legal enforcement issues. Getting farmers to sign licensing agreements seemed simply out of the question. To circumvent these hurdles, Monsanto devised another IP and marketing strategy, based on extensive licensing agreements with Indian seed companies.

In 1988, Monsanto and a well-established Indian seed company called Mahyco (Maharashtra Hybrid Seeds Company) formed a 50:50 joint venture, Mahyco Monsanto Biotech Limited (MMB), to market Bt cotton seeds in India.[3] In 1996, Mahyco imported 100 grams of Monsanto's Bt cotton seeds with the authorization of the Ministry of Environment and Forests.[4] Under a licensing agreement with Monsanto, MMB then used these donor seeds to "introgress" the Bt trait into its own hybrid cotton varieties. In 2002, MMB marketed the first three Bt cotton varieties—Bt Mech 12, Bt Mech 162, and Bt Mech 184. MMB also sublicensed the Bt gene to Indian seed companies. In 2004, the first variety developed by an Indian seed company under a sublicensing agreement with MMB was released (Rasi Seeds' RCH2). Around fifty Indian seed companies eventually became MMB sublicensees.

The sublicensing agreements between Monsanto and Indian seed companies remain private and confidential. However, as the agreements came under intense scrutiny, a number of provisions became public. Under the agreement, Monsanto supplies donor seeds incorporating the Bt trait to a seed company, which then transfers the Bt trait to its own varieties through conventional breeding techniques. Monsanto insisted that seed companies only introgress the Bt trait into proprietary hybrid cotton varieties, as opposed to open pollinated varieties (Sally and Singh 2019). As discussed in the introduction, seeds from hybrid varieties can be saved on-farm, but the yield declines after the first generation, thus creating an incentive for farmers to buy seeds every year. In other words, by restricting the Bt gene to proprietary hybrid varieties, Monsanto ensured that farmers would have to buy seeds in every planting season (Kranthi 2012).[5] The contract stipulated that seed companies had to make an upfront payment to Monsanto of 50 lakhs INR (in 2002, this represented roughly 100,000 USD).[6] In addition, seed companies were required to pay a recurring fee as a percentage of the value of each packet of seeds they sell. The price of a 450 g packet of BG-I cotton seeds (used to sow one acre) was initially between 1,600 and 1,800 INR (33–37 USD). Three-quarters of this amount, or 1,250 INR (26 USD), was passed on to Monsanto as royalties.

MMB holds the rights to the Bt technology, but seed companies can obtain plant breeders' rights over the Bt cotton varieties they develop under the Protection of Plant Varieties and Farmers' Rights (PPVFR) Act. This gives seed companies exclusive rights over the marketing of a Bt cotton variety. However, in order to obtain a plant breeder's certificate from the relevant public authority, a seed company was required, until 2017, to submit a No Objection Certificate issued by Monsanto. As its name implies, the certificate literally stipulated that Monsanto did not object to the registration of a plant variety developed by a seed company and containing Monsanto's Bt gene. This was a key element of the royalty collection system; it meant that Monsanto had the upper hand in negotiating sublicensing agreements with seed companies, as the latter depended on this certificate to be able to obtain plant breeders' rights over their Bt cotton varieties. In effect, Bt cotton is an example of a government agency advancing the interests of the industry it is charged with regulating instead of the public interest—a phenomenon known as regulatory capture. In India as in Brazil, the IP system implemented by Monsanto was based on the use of private contracts. However, in India, it can be characterized as a form of hybrid public–private arrangement because it relied to a certain extent on the cooperation of public agencies for its implementation.

In sum, under the sublicensing system implemented by Monsanto, Bt cotton seeds were sold either directly by MMB, Monsanto's joint venture and licensee, or by seed companies under sublicensing agreements with MMB.[7] As a result, by the early 2010s, Monsanto controlled, directly or indirectly, over 95 percent of the Indian Bt cotton market (Jayaraman 2012).

THE BT COTTON LEGAL DISPUTE

The conflict over Bt cotton seed prices and royalties erupted in the mid-2000s in the southern State of Andhra Pradesh, a major seed producer and cotton-growing state. By 2006, Bt cotton was cultivated on two-thirds of the cotton area in the state (GRAIN 2006). A 450 g packet of Bt cotton seeds was initially sold for 1,600 to 1,800 INR—that is, six times the price of a packet of conventional cotton. Yet in the first growing seasons, farmers in all three of India's major cotton-growing states (Maharashtra, Gujarat, and Andhra Pradesh) reported problems, ranging from the failure to germinate to pest attacks, fungal diseases, poor quality cotton, and even complete crop failure (Krishnakumar 2004).

In August 2005, two left-leaning farmers' organizations[8] filed a complaint against MMB before the Monopolies and Restrictive Trade Practices Commission. They alleged that MMB was charging exorbitant fees for seeds with the Bt trait and demanded that MMB be barred from setting the rate of royalties arbitrarily. In January 2006, the Andhra Pradesh (AP) government also asked the commission to open an investigation against MMB and some of its sublicensees,[9] arguing that MMB was engaging in restrictive trade practices and charging a higher rate of royalties in India than in China, where it was around 50 INR (1 USD) per packet—25 times less than the fee charged in India at the time (CCI 2016).[10] According to the State Minister of Agriculture, between 2002 and 2005, cotton farmers in Andhra Pradesh spent 130 crores INR (30 million USD) on Bt cotton seeds. Of this amount, 60 percent went to MMB as royalties (GRAIN 2006). While the complaint was before the commission, MMB reduced the rate of royalties by 30 percent (CCI 2016).

In a decision delivered in May 2006, the commission found that the fact that seed companies could only produce Bt cotton by entering into sublicensing agreements meant that there was zero competition and the amount MMB charged in royalties was excessively high. The commission directed MMB to bring down the value it charged as royalties to a reasonable level on par with the value charged in other countries. It also directed the government to set the maximum sale price of a packet of cotton seeds under the Essential Commodities Act (*Govt of AP v. MMB*, 2006). The AP government celebrated the court order as a significant victory for farmers in the first public interest lawsuit filed by a state government on behalf of farmers (Venkateshwarlu 2006).

That same month, the AP government set the price of Bt cotton seeds at 750 INR (17 USD) per packet for BG-I and at 925 INR (20 USD) for BG-II.[11] MMB immediately challenged both the government and the commission in India's Supreme Court, on the grounds that the former's move was illegal and that the latter had overstepped its jurisdiction and was not authorized to set prices (Mehta 2006). The Supreme Court agreed to hear the case but declined to suspend the AP government order.

Emboldened by the AP government's success, in June 2006 seven other cotton-growing states signed a Memorandum of Understanding calling for a common approach to the issue of Bt cotton royalties.[12] Denouncing "the exploitation of farmers in the garb of modern technology," they called for setting up a national Seed Price Regulatory Authority and adopting comprehensive seed legislation (Pantulu 2006). The states of Maharashtra, Gujarat, and Madhya Pradesh followed the lead of Andhra

Pradesh and introduced price controls for Bt cotton seeds under the Essential Commodities Act.[13]

While the cases were pending before the commission and the Supreme Court, MMB entered into a Settlement and Release of Claims Agreement with its sublicensees to readjust trait value to 150 INR (3.63 USD) for BG-I (CCI 2016).[14] With the maximum retail price of a packet of cotton seeds set at 750 INR, MMB could obviously no longer charge 1,250 INR in royalties. In light of these settlements, MMB withdrew its appeal to the Supreme Court in 2009.

Around 2010, the states that had passed Cotton Seeds Acts started to regulate not only seed prices but also the value of the royalties charged by MMB. Andhra Pradesh fixed the royalties for BG-I at 50 INR (1.09 USD) and for BG-II at 90 INR (1.97 USD) (Kurmanath 2010). Seed companies later claimed that MMB merely ignored these regulations and continued to charge higher royalties (CCI 2016).[15]

During this period, the Congress-led United Progressive Alliance (UPA) coalition was in government (2004–2014). The Central Government intervened on more than one occasion to undermine the states' efforts at regulation, for example, by removing cotton seeds from the Essential Commodities Act to undercut states' regulation of cotton seed prices under that legislation.[16] Following the election of the right-wing nationalist Bharatiya Janata Party (BJP) in the May 2014 general election, the conflict over Bt cotton took on a new course, abetted by the confluence of a personal feud and nationalist politics.

The Hindu ultranationalist movement is organized into volunteer organizations (Rashtriya Swayamsevak Sangh, RSS), some of whom oppose GMOs and multinational corporations.[17] They include the RSS farmers' wing (Bharatiya Kisan Sangh, BKS), which claims two million members and is committed to traditional agricultural knowledge and practices, as well as the Swadeshi Jagran Manch (SJM), which advocates self-reliance and is critical of foreign direct investment. While RSS organizations are broadly aligned with the BJP-led National Democratic Alliance (NDA) government, they have sometimes taken to the streets to denounce the government's policies considered to be "anti-farmer."

With the election of the BJP—the RSS's political arm—in 2014, RSS organizations such as BKS and SJM gained more leverage with the Central Government. As the BKS vice president observed, "In the previous regime we had to stand on the streets to launch anti-Monsanto protests. But with this government we can sit and talk in a room—it's because we all believe in the same agenda" (Bhardwaj, Jain, and Lasseter 2017). The BKS is particularly outspoken in its criticism of Monsanto. In the words of its vice president: "It is important for all of us to unite to wage a war against Monsanto" (Bhardwaj, Jain, and Lasseter 2017).

The dispute over Bt cotton has also been driven by a personal feud between Monsanto and Prabhakar Rao, the chief executive officer (CEO) of one of India's largest seed companies, Nuziveedu Seeds. In 2003, Rao approached the Genetic Engineering Approval Committee (GEAC) for permission to commercialize Bt cotton directly. GEAC declined, and Nuziveedu was forced to become a sublicensee of MMB (Interview #14B). To make matters worse, MMB refused in 2015 to grant a discount on royalties to Nuziveedu, jeopardizing Rao's plan to take his company public (Bhardwaj, Jain, and Lasseter 2017). At that point, Rao started lobbying the Central Government to intervene in the dispute. Rao was then president of the National Seed Association of India (NSAI), a seed industry body. In 2015, NSAI approached the Minister of Agriculture to initiate proceedings in the Competition Commission of India (CCI) to investigate whether MMB was abusing its dominant market position. An antitrust investigation was formally launched in February 2016.

The conflict between Monsanto and Nuziveedu came to a head in November 2015, when Monsanto declared Nuziveedu Seeds and two of its subsidiaries (Prabhat Agro Biotech and Pravardhan Seeds) to be in breach of payment obligations and terminated their sublicensing contract. According to Monsanto, Nuziveedu fell behind on royalty payments and owed Monsanto more than 20 million USD (Bhardwaj, Jain, and Lasseter 2017). Nuziveedu, for its part, argued that Monsanto had been illegally charging royalties above the state-stipulated rates.

In December 2015, the Ministry of Agriculture issued the Cotton Seeds Price (Control) Order, or CSP Order (GoI 2015). The order drew on the authority of the Central Government to control the prices of essential commodities in the public interest under the Essential Commodities Act.

It allowed the government, on the recommendations of a special committee, to set both the maximum price of Bt cotton seeds and the percentage of royalties that could be charged. While that Act had been used in the past to control the retail price of certain commodities, this was the first time it was used to regulate royalties. Monsanto immediately challenged the CSP Order before the Delhi High Court on the grounds that it was illegal as well as unconstitutional (Bera and Sen 2016).

In February 2016, Monsanto filed another lawsuit in the same court, this time against Nuziveedu for patent infringement. Nuziveedu, Monsanto claimed, had illegally continued to use the Bt technology after the termination of its sublicensing agreement.

The same month, the Competition Commission issued a preliminary report on its antitrust investigation on MMB (CCI 2016). The commission found that, on initial examination, sufficient evidence supported the case that MMB had violated the Competition Act (GoI 2002b). The commission noted that Monsanto was indeed in a dominant position since it was the sole provider of the BG-II Bt cotton technology, used on 99 percent of the area under Bt cotton cultivation in India. It found that the conditions for the termination of the sublicensing agreement were stringent and even abusive. It took issue, for example, with a provision of the 2015 agreement stipulating that MMB could terminate a sublicensing agreement if the central or state governments passed regulations on trait fees.[18] The commission also questioned the economic justification of calculating the royalties based on the maximum retail price of cotton seeds, since the Bt trait was only one among many factors (genetic composition, climatic conditions, and others) contributing to its performance. The order instructed the Director General to conduct a full investigation. Monsanto challenged the CCI order before the Delhi High Court, arguing that the commission lacked the authority to examine issues pertaining to intellectual property and trademarks. The Court allowed the CCI to pursue its investigation but added that any CCI order would only be given effect with the leave of the Court.

In March 2016, Monsanto threatened to depart India altogether. Its India CEO declared: "It will be difficult for [MMB] to justify bringing new technologies into India in an environment where such arbitrary and potentially destructive government interventions make it impossible to recoup research and development investments focused on delivering extensive farmer benefits and where sanctity of contracts is absent" (*The Hindu* 2016). According to a competition lawyer, it was a mistake for Monsanto to threaten to reevaluate its position in India, as these threats did not go down well with the government. In fact, they had the opposite effect (Interview #4A). Four days later, the Ministry of Agriculture, under the CSP Order and on the recommendation of the special committee, fixed the price of cotton seeds at 800 INR (12.50 USD). The Ministry of Agriculture canceled royalties for BG-I and imposed a sharp 74 percent decrease in royalties for BG-II.

The same month, India's Department of Industrial Policy and Promotion served a show-cause notice to Monsanto, calling on it to explain why its patent on BG-II should not be revoked, given that it had lost its effectiveness against the pink bollworm (Deshpande 2016).[19] The department was acting on a request of the BJP's farmers' wing (Kisan Morcha) to the Ministry of Agriculture (Kurmanath 2016).

In May 2016, in a bold move, the Ministry of Agriculture issued the Draft Licensing Guidelines and Formats for GM Technology Agreements, a compulsory licensing regime for GM technologies. These guidelines stipulated that a patent holder could not refuse to grant a license to any eligible seed company. If the patent holder failed to do so, the licensee automatically obtained a license under the Fair, Reasonable, and Non-Discriminatory terms described in the guidelines.[20] In practice, this meant that Monsanto would lose control both over who becomes a sublicensee and over the terms of licensing.

However, within two days of the Draft Licensing Guidelines' publication, the Indian government backtracked and announced that it would invite public comments on the document for three months. It was later revealed that the US ambassador had intervened directly with the Indian government to have the guidelines withdrawn (Bhardwaj, Jain, and Lasseter 2017). The public review was held between June and August 2016, but the guidelines were not reinstated.

In August 2016, citing "the uncertainty in the business and regulatory environment," Monsanto announced it had withdrawn its application in India for the next generation of Bt cotton, a stacked Bt and Roundup Ready variety called BG-II Roundup Ready Flex (Bhardwaj 2016). The announcement letter referred explicitly to the Licensing Guidelines as having "alarmed us and raised serious concerns about the protection of intellectual property rights" (Bhardwaj 2016).

In June 2017, in another blow to Monsanto, the Ministry of Agriculture canceled the requirement to submit a No Objection Certificate issued by the patent holder for the registration of a hybrid variety containing a patented trait (Fernandes 2017). Seed companies had been lobbying the government to cancel the No Objection Certificate requirement that gave Monsanto the upper hand in negotiating sublicensing agreements. The companies argued that there was no basis in the Protection of Plant Varieties and Farmers' Rights Act (PPVFR Act) for such a requirement, and that the Act only required a declaration from the applicant that it had legally acquired the genetic material or parental material used to develop a new plant variety.

The Delhi High Court delivered two decisions in the patent infringement lawsuit that Monsanto filed against Nuziveedu Seeds in 2016.

In the first decision, delivered in March 2017, Judge Gauba ruled that Nuziveedu's demand to renegotiate royalties in accordance with the government's CSP order was legitimate. Given Monsanto's refusal to do so, its decision to terminate its sublicensing agreement with Nuziveedu was illegal. The judge ordered Monsanto to restore the contract and abide by the trait fee fixed by the government under the CSP Order. The judge also stated that while he was not in a position to rule on the complex issue of patent validity, he found Nuziveedu's argument that Monsanto's patent on Bt cotton had been wrongly granted by the Indian Patent Office "*prima facie* to be devoid of merit" (*Monsanto v. Nuziveedu*, 2017).

Both parties appealed before the same court to contest specific aspects of the decision. The parties also agreed that the Court would decide the issue of patent validity on the basis of the evidence already submitted to the court. For Monsanto to forfeit its right to a full trial on such a fundamental issue was, as one IP lawyer put it, either "incredibly brave or incredibly overconfident" (Reddy 2018a).

The second decision in the case was issued a year later, in April 2018. In a landmark ruling, Justices Ravindra Bhat and Yogesh Khanna ruled that Bt cotton seeds were *not* patentable in India, in effect revoking

Monsanto's patent. This decision was significant because it was the first time that a court examined the legality of patents on biotech traits under Indian law. In this case, the patent in question was no. 214436, one of two patents obtained by Monsanto in India on a method for producing Bt plants (Corbin and Romano 2008). First, the justices reasoned that the narrowing of the claims in the patent application so as to conform to the Indian legislation had important implications for the scope of the resulting patent. Second, they were of the opinion that the subject matter of the patent—nucleic acid sequences—did not qualify as microorganisms patentable under Article 27.3(b) of the Agreement on Trade-Related Aspects of Intellectual Property Rights (TRIPS). Third, the justices maintained that the transfer of the Bt trait to plant varieties through hybridization was an essentially biological process, exempted from patentability under Section 3(j) of the Indian Patents Act. Fourth, they argued that "use" of the patented invention could not be construed to include use of the plants and their offspring, both of which are explicitly excluded from patentability under Section 3(j) of the Patents Act. Finally, the justices also pointed out that Indian legislation guarantees substantive rights to farmers.[21]

Monsanto appealed the Delhi High Court ruling before the Supreme Court, which ruled in January 2019 that the Delhi court could not invalidate the patent without conducting a full-fledged trial—and then sent the case back to that court for reexamination (*Monsanto v. Nuziveedu*, 2019).

In a reversal of matters, Bayer AG—which acquired Monsanto in 2018—announced in April 2021 that it had reached a legal settlement with Nuziveedu that would end all ongoing litigation, including the Delhi High Court infringement lawsuit (Bhardwaj and Kaira 2021). Under this settlement, biotech companies and domestic seed companies agreed on a "framework on trait value and licensing agreements" fixing trait fees at 5 to 20 percent of seed value (Kurmanath 2021). The out-of-court settlement put an end to the 16-year conflict between Monsanto and domestic seed companies. Unfortunately, the settlement also prevented the Delhi High Court—and eventually the Supreme Court—from ruling on the all-important question of the patentability of plant-related inventions under Indian law.

MONSANTO'S INTELLECTUAL PROPERTY RIGHTS TO BT COTTON

In 1994, the Indian government rescinded a patent granted by the country's Patent Office to the US company Agracetus on a method for producing genetically engineered cotton.[22] The method consisted of the use of a bacterium (*Agrobacterium tumefaciens*) to "ferry" foreign genes into cotton cells—one of the most common methods used to genetically engineer plant cells. This patent was extremely broad and covered all transgenic cotton plants transformed using this method. In 1991, Monsanto licensed from Agracetus the right to use this technology to genetically modify plants. In 1996, Monsanto acquired Agracetus, and with it all the company's patents on genetically engineered cotton and soybeans.

The Indian government had been tipped off about this patent by a communiqué of the Canadian-based NGO Rural Advancement Foundation International (RAFI, now ETC Group), denouncing the unprecedented broad scope of the patent, which would give Agracetus monopoly control over virtually all transgenic cotton plants and seeds until 2008 (RAFI 1993). The cancellation of Agracetus's patent marked one of only two instances in which the Indian government has revoked a patent in the public interest, citing its far-reaching implications for India's cotton economy as well as its negative impact on farmers.[23]

When the patent was originally granted to Agracetus in 1991, agriculture and intellectual property were under discussion in the Uruguay Round of GATT negotiations (1986–1994), but the TRIPS Agreement was not yet on the horizon. The 1970 Indian Patents Act explicitly excluded agriculture and horticultural methods of production from patentability. As Anumita Roychowdhury reported at the time, "Embarrassed [Department of Industrial Development] officials are on the defensive, saying, 'The area of biotechnology is relatively new and a very complex one. Interpreting [the Indian Patents Act] can be confusing because it is not explicit on microbiological processes. The interpretation of these have [*sic*] been left open even in the Uruguay round of GATT'" (Roychowdhury 1994).

By the time MMB introduced BG-I Bt cotton in 2002, India was in the process of revising its Patents Act to conform to the TRIPS Agreement. Monsanto's patent rights over Bollgard-I in India, like those over Roundup

Seeds of doubt: Monsanto never had Bt cotton patent

TNN | Jun 8, 2015, 02:17 AM IST

NAGPUR: Fourteen years after US multinational Monsanto brought the genetically modified (GM) Bt Cotton (Bollgard) to India, there is no clarity on the discovery having ever been patented in the country. Clueless Indian farmers and seed manufacturers have paid crores as royalty to the company from 2002 until 2006, when the company came out with Bollgard 2, which was, incidentally, patented.

Two arms of the central government differ on the patent issue. The Central Institute of Cotton Research (CICR), in an RTI reply to farm activist Vijay Jawandhia, emphatically stated that Monsanto's 'crylac Mon 531' gene was never patented in India. However, the ministry of environment and forests (MoEF) wrote to him that the Bt seed developed by University of Agriculture Sciences (Dharwad), which was found to contain the Mon 531 strain, "cannot be launched in the market" due to a "patent violation". It did not specify who held the patent.

Queries to Monsanto specifically on the patent issue were avoided. "Monsanto has proprietary rights in its regulatory data as well as its biological materials, trade secrets and know-how, which are also protected under Indian law. The Mon 531 is subject to such rights," said a company spokesperson and never got back on a query seeking the patent number.

3.1 "Monsanto never had Bt cotton patent," *Times of India*, June 2015

Ready soybean in Brazil, were shrouded in confusion, and it was widely believed that Monsanto held a patent on BG-I.[24] In 2015, thirteen years after the introduction of Bt cotton, the *Times of India* broke the story that Monsanto had in fact never held a patent on first-generation Bt cotton (Arya and Shrivastav 2015).

According to the *Times* journalists, a farm activist who filed a Right to Information request about Monsanto's patent rights received contradicting answers from two different government bodies. The Central Institute of Cotton Research (CICR) responded that Monsanto's Cry1Ac gene (commercially known as BG-I Bt cotton) had in fact never been patented in India. However, the Ministry of Environment and Forests said that the Bt cotton developed by CICR using that same gene could not be put on the market because it would infringe on Monsanto's patent. The CICR, it turns out, was right: Monsanto had never obtained a patent for Bollgard-I Bt cotton in India. While some people knew this,[25] it came as a surprise to many.

According to the farm activist who filed a Right to Information request, "Without the active support of ministry officials, it would have not been possible to keep the fact that Monsanto has no patent on MON 531 [BG-I] under wraps" for so long (Arya and Shrivastav 2015). Another activist told me that she believes that in the early years there was an "unwritten etiquette"—a tacit understanding that government regulators and private companies would respect patent rights granted in other jurisdictions (Interview #14B). This meant, for instance, that they would recognize Monsanto's IP rights over BG-I, even though it did not have a patent in India. According to the same activist, this stemmed, in part, from the government's desire to maintain good relations with the company because it saw access to the technology as being in the country's best economic interest (Interview #14B; see, also, Newell 2007).

Monsanto most likely did not apply for a patent for BG-I in India in the early 2000s because the process of amending the Patents Act had not yet been completed. As we saw in chapter 1, key amendments to bring the Patents Act into line with the TRIPS Agreement were passed in 1999, 2002, and 2005. Moreover, it takes some time for patent offices to adjust their examination guidelines to integrate such major legislative changes. Although these changes were in the making, the Indian Patent Office would have been unlikely to grant a patent on anything related to a plant at the time (Interview #67).

In the absence of a patent, Monsanto "creatively" leveraged its control over the biosafety approval process to negotiate sublicensing agreements.[26] In a 2006 report, the World Bank (2006, 33) admitted to this, stating that "As the company does not 'own' the gene in India, the contract is based on access to the biosafety data that are necessary for approving any transgenic variety." This corroborates Diego Silva's argument that biosafety narratives, regulations, and practices have been mobilized at different scales as instruments for the enforcement of IP rights (Silva 2017).

A spokesperson for MMB explained Monsanto's IP rights over BG-I Bt cotton in India as follows:

Monsanto has various intellectual property rights covering its cotton technologies in India. The MON531 event (BG-I) which expresses the Cry1Ac gene has not been, in itself, patented in India by Monsanto. . . . However, Monsanto does enjoy proprietary rights over the MON531 event pertaining to regulatory data, biological materials, trade secrets, know-how and the like. The technology registrant also has proprietary rights and corresponding obligations over technologies commercialized under its registration/approvals for MON531 granted by regulatory authorities such as GEAC and applicable Indian laws.[27] (Kaveri Seeds 2015)

However, according to one IP legal expert, this statement is not legally tenable. The expert gives as examples the vagueness of the wording ("and the like") and the fact that "proprietary rights"—the protection of trade secrets or commercial information that is privileged or confidential—is a US concept not recognized in Indian law (Interview #42).

By the time Monsanto introduced BG-II in 2006, India had completed the process of amending the Patents Act to comply with the WTO TRIPS Agreement. Monsanto obtained two patents related to BG-II. The first is a broad patent on Bt technology (no. 214436—Methods for transforming plants to express bacillus thuringiensis delta endotoxins) granted by the Indian Patent Office in 2008 and valid until 2019 (Corbin and Romano 2008).[28] The second is a patent specific to BG-II technology (no. 232681—Cotton Event MON 15985 and compositions and methods of detection), granted by the Indian Patent Office in 2009 and valid until 2022 (Shappley et al. 2009).

The original application for Patent no. 214436 included 58 claims covering "a plant," "a progeny plant," "a plant cell," and "a plant tissue," among others. The Patent Office objected to all but three product claims that concerned nucleic acid sequences (*Nuziveedu v. Monsanto*, 2018). In much the same way as the Brazilian Roundup Ready soybean patent was obtained, the Bt cotton patent finally granted in India was modified to focus on the *process*, since most of the *product* claims were not admissible under the Patents Act. As a result, the Indian patent contains 24 process claims (for example, "A method for producing a transgenic plant . . .") as well as three claims related to nucleic acid sequences (claims 25–27) (Corbin and Romano 2008).

According to the Monsanto spokesperson quoted above: "Monsanto has been granted a patent in respect of the MON 15985 event [BG-II] in India. We claim proprietary rights over the MON 531 event [BG-I] by virtue of the various buckets of rights available to us and hence *our claim has to be viewed from a holistic perspective rather than a singular patent lens*" (Kaveri Seeds 2015, emphasis added). The last sentence is revealing of the fact that Monsanto was operating on the principle that it should enjoy the same level of IP protection *as if* it had an Indian patent on BG-I, even though it did not.

Monsanto does not make public how much it earns in royalties. A company spokesperson told a journalist that under its contractual obligations, it cannot share competitive information such as royalty fees (Jishnu 2010a). Nor does Monsanto reveal how it determines the amount of royalties it charges (Jishnu 2010b, 2010c). The investigation by India's

Monopolies and Restrictive Trade Practices Commission concluded that "Monsanto was in a position to charge arbitrarily for the Bt cotton technology and could not offer any rational explanation for arriving at the trait value of Rs.1250 per packet" (CCI 2016, 7).

Pressed to explain, the company has put forward different explanations about its royalty fees over the years. In 2010, a top representative of Monsanto told a journalist that "the trait value [royalties] charged is relative to the additional income that farmers earn from Bt seeds, a formula that includes the savings in pesticide usage" (Jishnu 2010b). This explanation is problematic for a number of reasons: there are no reliable ways to calculate potential savings, and calculations vary greatly from region to region. Moreover, results depend on a wide range of factors other than the Bt trait, including the genetic makeup of the seed (the Bt trait is only one component), weather, growing conditions, and so on. As critics point out, if this is the case, how can Monsanto continue to earn royalties in the event of crop failure? It is as if success is always attributable to Monsanto, whereas failure is the making of farmers or nature itself (Jishnu 2010b).

The dispute over Bt cotton seed prices and royalties offers a compelling window into the global and national politics of intellectual property. In the early 2000s, an "unwritten etiquette" ensured that the Indian government would proceed as if Monsanto had a patent on BG-I—when it actually did not. Corporate lobbying certainly played a part in this tacit agreement, as did the government's belief that it could not afford to miss out on this technology and India's desire to be a global player in the biotech industry (Newell 2008). In practice, this meant that in India Monsanto had a free hand, as it did in Brazil, to implement a private royalty collection system based on extensive sublicensing agreements with Indian seed companies, and to extract extremely high rates of royalties. This system relied on the cooperation of the state for its implementation, as witness the No Objection Certificate—a clear private interference into government regulation of plant breeders' rights. Another example of this reliance is the fact that the government did not allow the commercialization of Bt cotton varieties developed in the public sector, nor did it allow Nuziveedu to commercialize Bt cotton directly. At the state level, farmers'

Cotton Crunch

Seed giant Monsanto meets its match as Hindu nationalists assert power in Modi's India

3.2 "Seed giant Monsanto meets its match as Hindu nationalists assert power in Modi's India," *Reuters*, March 2017

organizations successfully mobilized to get state governments to regulate royalties and seed prices. Yet Monsanto's royalty collection system was only seriously challenged following the election of the BJP, when Indian seed companies and RSS groups gained more influence with the Central Government. The shift in power was captured well in the title of a newspaper article: "Seed giant Monsanto meets its match as Hindu nationalists assert power in Modi's India" (Bhardwaj, Jain, and Lasseter 2017).

The state, of course, is not monolithic, and government bodies have been rife with internal tensions over the governance of agricultural biotechnology. India's Prime Minister Narendra Modi, in spite of his high-pitched nationalist rhetoric, has eased restrictions on foreign direct investment and openly supported multinational corporations entering and operating in India. His government has reiterated its public commitment to respecting international IP standards and has been reluctant to issue compulsory licenses for drugs (Damodaran 2016). It has also come out in support of GM crops, echoing the biotech industry's argument—amply refuted—that these crops will bring about a much-needed increase in food production (Sehgal 2015).[29] On the Bt cotton issue, however, the RSS ultranationalist views against the entry of foreign corporations into the country seem to have prevailed so far (Kang 2016; Andersen and Damle 2019). According to one activist, this very public round of regulation of royalties is a "win–win" for the BJP-NDA government: in the context of the ongoing agrarian crisis and amid all the other pro-corporate policies it has been promoting, it gives the impression that the government is not beholden to large corporations (Interview #14B).

In a similar way to what happened in Brazil, what started out in India as a dispute over seed prices and royalties evolved into a legal challenge to the patents themselves. The patent infringement case that Monsanto brought against its sublicensee has for the first time prompted the Indian judiciary to consider the very legality of biotech seed patents under the Indian legislation. Unfortunately, the out-of-court settlement reached between Bayer-Monsanto and Nuziveedu in 2021 ended all ongoing litigation—ensuring that the intellectual property aspects of agricultural biotechnology would remain a gray area in India, at least for the foreseeable future.

4

WHO OWNS BT BRINJAL?

Oh, you know, law is only an instrument. You know how things work in the country, there are so many laws which we don't apply.

—High-ranking government official, in response to the Bt brinjal complaint[1]

Like Bt cotton, Bt brinjal is genetically engineered by inserting a gene from the soil bacterium *Bacillus thuringensis* into the eggplant's genome (eggplant is known as *brinjal* in India). The Bt gene enables plant cells to produce a protein, Cry1Ac, that behaves as a toxin against the shoot and fruit borer, a pest that affects eggplant. When a shoot and fruit borer larva feeds on Bt brinjal, it ingests the Bt toxin, which perforates the gut and causes death (PANAP 2012). The Bt gene is patented as Event EE-I.[2]

Mahyco, the Indian seed company that became Monsanto's partner for the marketing of Bt cotton (see chapter 3), started to work on the development of Bt brinjal around 2000. In 2006, Mahyco filed a patent application at the Delhi Patent Office for a "Transgenic Brinjal (*Solanum Melongena*) Comprising EE-1 Event." The same year, Mahyco submitted biosafety and efficacy data to the Indian regulatory agency—the Genetic Engineering Approval Committee, GEAC—and applied for permission to conduct large-scale trials (Shah 2011).

When Bt brinjal came up for approval, Bt cotton was already widely cultivated in India. However, as the first genetically engineered *food* crop,

Bt brinjal took on a special significance. Along with the potato, eggplant is one of the most important vegetables produced and consumed in India. Over 2,500 eggplant varieties are in cultivation, and the 9.5 million tons of eggplant produced annually is mostly consumed domestically. Brinjal is grown mainly in the three eastern states of Bihar, Orissa, and West Bengal, with a smaller but nonetheless significant production elsewhere in India. Most farmers who grow eggplants are small-scale (Andow 2010; PANAP 2012).

The imminent commercial approval of Bt brinjal caused a heated controversy, and a public interest lawsuit in the Supreme Court of India led to a ban on ongoing field trials.[3] The regulatory agency set up two expert committees to examine the biosafety of Bt brinjal and eventually decided to give the green light to Bt brinjal. This decision prompted the Environment Minister to launch a nationwide public consultation that culminated in a national moratorium in February 2010 (Chowdhury and Srivastava 2010). The biopiracy accusations related to Bt brinjal discussed in this chapter arose in the context of this public consultation. However, to understand the lawsuit, it is necessary to first step back and examine how Indo–US engagement on agriculture and intellectual property (IP) policy in the 1990s and 2000s led to the development of Bt brinjal.

A GLOBAL PUBLIC–PRIVATE PARTNERSHIP: THE ABSP-II PROJECT

In 1990, the US National Research Council published a report commissioned by the US Agency for International Development (USAID) titled *Plant Biotechnology Research for Developing Countries*. The report emphasized "new ways of doing business" and the authors' belief that "the new actor is the private sector" (NRC 1990, 5–7). The report reveals a curious tension between perspectives on the needs of developing countries and on their potential as untapped markets. The authors of the report argued, for example, that *Bacillus thuringiensis* "offer great potential benefits for the less developed countries" and, almost in the same breath, that "Third World countries offer a vast untapped market for both the use and development of novel Bt products" (NRC 1990, 21). They concluded that "there is potential for joint ventures and transfer of technology between the private sector in industrialized countries and the [least developed countries] that have developed some expertise in Bt research" (NRC 1990, 21). The authors adopted a distinctively paternalistic tone regarding the role of the United States in the formulation of public policy in these countries. In the area of biosafety, for example, they wrote: "Developing countries could modify US standards to fit their needs, rather than starting from scratch. But they need *objective, authoritative* advice. Many countries have difficulty in deciding which products to license and which companies to allow to develop and test products, and as a result, err on the side of caution, so that the use of safe products is not being permitted. This suggests an important role for AID's technical assistance through USAID missions and regional programs" (NRC 1990, 14–5, emphasis added).

One year after the report, these recommendations were followed up by the creation of the Agricultural Biotechnology Support Project (ABSP-I). That project was funded by USAID and was based at the Institute of International Agriculture at Michigan State University. It brought together US agricultural universities, private seed companies, and international agricultural research centers. Its final report, published in 2003, concluded that public–private sector partnerships were critical, but deplored the fact that the project had produced few results in terms of bringing transgenic products to the market (ABSP 2003). One of its key recommendations was to focus on product development instead of upstream research, more specifically on the development and use of transgenic technology (Kent 2007).

That same year, ABSP-I was replaced by ABSP-II—a consortium of private and public institutions also funded by USAID but now led by Cornell University. The university has played a leading role in the development of agricultural biotechnology. In the late 1980s, scientists at Cornell developed one of the most commonly used methods to induce genetic transformation in plants: biolistics, commonly known as the gene gun.[4] The ABSP-II consortium includes over fifty governmental agencies, national and international agricultural research centers, universities, and private companies; it has three regional centers, in East Africa, Southeast Asia, and South Asia (ABSP-II n.d.). The stated objective of ABSP-II is: "The safe and effective development and commercialization of bio-engineered products as a complement to traditional and organic agricultural approaches in developing countries, and . . . to make such products available to farmers in forms they can use, to help reduce poverty and hunger and to boost food security, economic growth, environmental quality and nutrition" (Sathguru 2013).

In India, ABSP-II's main project has been the development and commercialization of Bt brinjal.[5] The public institutions in India that are part of the consortium are the Department of Biotechnology, which is under the Ministry of Science and Technology and is responsible for promoting and regulating biotechnology;[6] and the National Bureau of Plant Genetic Resources (NPBGR), which is part of the Indian Council of Agricultural Research (ICAR) and is responsible for the ex situ management of plant genetic resources.[7] The ABSP-II consortium also includes three public agricultural institutions: the University of Agricultural Sciences in Dharwad, Karnataka (UAS Dharwad); the Tamil Nadu Agricultural University (TNAU); and the Indian Institute of Vegetable Research (IIVR), in Uttar Pradesh. The two latter institutions were established in the early 1970s and were an integral part of Green Revolution efforts to set up an Indian agricultural research and extension system modeled on the land-grant universities in the United States. As for UAS Dharwad, it was established more recently, in 1986.[8] On the private side, the ABSP-II consortium includes Mahyco (Monsanto's partner in India) and Sathguru Management Consultants. Sathguru is a private consulting firm based in Hyderabad, Andhra Pradesh. As regional manager for ABSP-II's South Asia center, Sathguru acts as the Indian coordinator for the Bt brinjal project (ABSP-II n.d.).

ABSP-II was part of the Knowledge Initiative on Agriculture (KIA) signed in July 2005 by India and the United States.[9] KIA board members included prominent US-based multinational corporations in the agricultural and food processing, trading, and retailing sectors, including Monsanto, Archer Daniels Midland, and Walmart. The ABSP-II consortium also worked closely with the International Service for the Acquisition of Agri-Biotech Applications (ISAAA), an industry group that promotes biotech crops globally, notably in the Global South.[10] The following quote by ISAAA president Clive James conveys the spirit of philanthrocapitalism behind the ABSP-II project:

In the spirit of sharing and caring, the Bt brinjal technology used for hybrids has been generously donated by its developer, M/s Maharashtra Hybrid Seeds

Company (Mahyco) to public institutes in India, Bangladesh and the Philippines for use in open-pollinated varieties of brinjal, in order to meet the specific needs of small resource-poor farmers in neighboring countries in the region where brinjal is an important crop. This is an excellent working example of a model philanthropic public/private sector partnership that has facilitated the generous donation of a biotechnology application by a private sector company for use by public sector institutes to meet the needs of small resource-poor farmers. (James, quoted in Choudhary and Gaur 2009, iii)

James's description of Bt brinjal as meeting the needs of small, resource-poor farmers in South and Southeast Asia stands in stark contrast to the comment made by India's Environment Minister, Jairam Ramesh, in his moratorium decision on the commercialization of Bt brinjal: "There does not seem to be any over-riding food security, production shortage or farmer distress arguments favouring the enormous priority that has been accorded to [Bt brinjal] by private companies, other than the well-known argument on the need to reduce pesticide use" (MoEF 2010a, 3).

THE BIOPIRACY PUBLIC INTEREST LAWSUIT

In February 2010, the last of the public meetings on the commercial release of Bt brinjal organized by the Ministry of Environment was held in Bangalore, the capital of Karnataka, in southern India. When perusing the documents made available in the context of the public consultations, a local NGO called Environment Support Group (ESG)[11] discovered that the developers of Bt brinjal had not applied to the National Biodiversity Authority (NBA) for permission to access local eggplant varieties, as required by the Biological Diversity Act.

The Act was enacted in 2002 to meet India's obligations under the UN Convention on Biological Diversity, notably in terms of both biodiversity conservation and equitable benefit-sharing arising out of the use of traditional biological resources and knowledge (GoI 2002a). Under the Act, Indian biodiversity authorities must authorize access to biological resources and associated knowledge for either research or commercial use. Indian citizens and companies must obtain permission from state biodiversity boards, while foreigners and foreign companies must obtain permission from NBA. Authorization from NBA is also required to apply for any IP rights in India or abroad on biological resources originating in

India. Infringement of the Biological Diversity Act is liable to fines and imprisonment.

At the February 2010 meeting in Bangalore, ESG submitted a petition explaining that it believed the process by which Monsanto-Mahyco and its collaborators had accessed and genetically engineered local eggplant varieties was in violation of the Biological Diversity Act (ESG 2010a). Their concerns were not initially taken seriously by the Ministry of Environment and Forests (MoEF). In a footnote to the February 9 decision introducing a national moratorium on the commercialization of Bt brinjal, the questions raised by ESG were dismissed as a "wholly unjustified controversy" (MoEF 2010a, 2).

Unfazed, ESG approached the Karnataka State Biodiversity Board, or KBB (ESG 2010b). According to ESG, "the Board rigorously investigated the case, issued notices on all accused institutions, conducted workshops and hearings, visited the [University of Agricultural Sciences in Dharwad] to investigate its role, and repeatedly sought advice from NBA on how to proceed, considering that foreign companies were involved" (Saldanha and Rao 2011, 27). In March 2010, KBB communicated its preliminary findings to NBA, stating that there was strong evidence to support the contentions of biopiracy that ESG had raised. The KBB confirmed that the local varieties had been used without the clearance of local, state, or national authorities, and charged that this amounted to a clear violation of the Biological Diversity Act (KBB 2010). In a direct reference to the international legal framework on farmers' rights, in particular the FAO Plant Treaty,[12] KBB also stated that "This action denies the local communities, who have cultivated and protected these varieties from time immemorial, their due right to benefit from the commercial gains that would be made from the access and use of these biological resources" (KBB 2010, 138).

A year later, in May 2011, KBB submitted another report to NBA in which it provided details of the agreement for the development of Bt brinjal. According to the report, the director of research at UAS Dharwad stated in a letter that the Bt brinjal technology made available to Mahyco had been transferred to UAS Dharwad under ABSP-II, and that UAS Dharwad had used six local varieties for the development of Bt brinjal without prior approval from either the State Biodiversity Board or the National Biodiversity Authority (KBB 2011).

During its board meeting the following month, NBA decided to take action. According to the minutes of the meeting, it was decided that NBA would proceed legally against Mahyco, Monsanto, and all others concerned to take the issue to its logical conclusion (NBA 2011a). This decision, however, was only made public by the new Environment Minister, Jayanti Natarajan,[13] in September 2011 (Press Information Bureau 2011). In spite of this announcement, the minutes of an NBA board meeting in November 2011 revealed a change of heart. During that meeting, three board members suggested that no legal action be taken since the issue was merely a research collaboration that was exempted under the purview of the Biological Diversity Act (NBA 2011b).

The board members were referring to a provision of the Biological Diversity Act that states that the authorization requirement is waived for collaborative research projects involving the transfer or exchange of biological resources or knowledge between Indian and foreign institutions, provided that they satisfy certain conditions (MoEF 2006). According to this line of interpretation, Bt brinjal, as part of the ABSP-II project, would be exempt from the purview of the Biological Diversity Act. As one legal scholar argues, "Section 5(3)(b) of the Biological Diversity Act is pretty clear that permission from the NBA is not required in collaborative research projects between government sponsored institutions of India and other countries if such projects have been approved by the Central Government. UAS was a government university. The entire ABSP-II project is a part of the Knowledge Initiative on Agriculture (KIA) entered into between the Governments of India and the US" (Reddy 2012).

At the November 2011 NBA meeting, there was also a clear will on the part of the chair to dissociate the accusations of biopiracy from the issue of biotechnology: "The Chairman has informed the members that the issue at hand is purely that of possible misappropriation of local brinjal varieties and has nothing to do with biotechnology per se and/or its application" (NBA 2011b, 11).

Despite its initial responsiveness to ESG's complaint, the Karnataka Biodiversity Board decided in January 2012 that the subject fell within the purview of the National Biodiversity Authority (KBB 2012). In a meeting the following month, NBA decided to initiate legal proceedings by a close vote of three to two (NBA 2012).[14] NBA then proceeded to document the case and prepare the criminal complaint. In November 2012, in yet another reversal, the two deputy conservators of forests who acted as co-complainants along with NBA officers were issued transfer orders with immediate effect, only days before they were due to file the complaint in the Karnataka High Court (Sood 2013). The KBB member secretary, however, took it upon himself to delay their transfer long enough to allow them to file the case in court (ESG 2013).

These delays and setbacks, together with the lack of clear political will to pursue the case on the part of the public authorities, prompted ESG to file a public interest lawsuit in the High Court of Karnataka in November 2012 (*ESG v. NBA*, 2012). The respondents were NBA, the Ministry of Environment and Forests (MoEF), KBB, the State of Karnataka, and the Union of India.

The objective of the lawsuit was to draw attention to what ESG argued was the ineffective implementation of the Biological Diversity Act. ESG pointed to NBA's failure to issue important regulations on access to biodiversity, the transfer of research results, and intellectual property, as well as its failure to recruit or hire an adequate number of taxonomists and to set up a permanent legal cell. ESG also denounced NBA's failure to take action against violations of the Biological Diversity Act.

Several public reports and audits published during this period lent weight to ESG's claims. In a 2010 audit report on the MoEF and NBA, the Comptroller and Auditor General (CAG) of India observed that "even six years after its formation, NBA could not notify important regulations like access to biodiversity, transfer of results of research and intellectual property rights, etc." (CAG 2010, 29). The CAG also noted that issuing approvals in the absence of a regulatory framework was highly problematic.[15] In 2012, the Agriculture Committee of the Lok Sabha released a report on the cultivation of genetically modified food in which it addressed the complaint made by ESG:

The Committee are not at all convinced by the dilatory response of NBA on this sensitive issue. The matter is very simple as to whether the Company in question has obtained any local biological resource for and in connection with development of Bt. Brinjal without prior approval of NBA and violated Section 3 of Biological Diversity Act, 2002. Taking so long in coming to a conclusion on this simple issue shows the NBA in a very poor light. It would also be worth mentioning here that during this period Chairman, GEAC was simultaneously also holding the charge of Chairman, NBA, from 11 November, 2010 to 11 August, 2011. The Committee not only desire a thorough inquiry in the matter of continued paralysis in decision making on a case of this dimension but also recommend that the NBA should decide upon this case without any further delay. (Lok Sabha 2012b, 281–82)

Another issue raised by ESG in the public interest lawsuit concerns the inclusion of threatened and endangered plant species in the list of Normally Traded Commodities. Section 40 of the Biological Diversity Act provides that "the Central Government may, in consultation with the National Biodiversity Authority, by notification in the Official Gazette, declare that the provisions of this Act shall not apply to any items, including biological resources normally traded as commodities." This was done via a Gazette Notification issued by the MoEF on October 26, 2009, in consultation with NBA (MoEF 2009). The notification included a list of 190 species which are normally traded as commodities. The rationale behind the list was to facilitate the trade of items, including biological resources, that are normally traded as commodities by exempting them from the purview of the Act. However, ESG argues that some fifteen plants that had been put on the list were either threatened, endangered, or critically endangered (*ESG v. NBA*, 2012, 21).

Solanum melongena—the botanical name of eggplant—was included in the list of Normally Traded Commodities (MoEF 2009). This led some to argue that eggplant was thus exempted by the government of India from the provisions of the Biological Diversity Act (Rao 2013). This interpretation, however, was refuted by the MoEF itself in a note of clarification published in February 2010:

Some concerns have been raised on the implication of this Notification. It is wrong to say that "190 plants have been dropped from the protection of the Biological Act, 2002." It is clarified *categorically* that this Notification applies only to *export* of these 190 items and these exports would not require prior permission of the National Biodiversity Authority. *However, it is also made clear that if these biological resources are used as a source for research or industrial purposes, they will require previous approval of the National Biodiversity Authority as per relevant provisions of the Biological Diversity Act, 2002.* (MoEF 2010b)

In the public interest lawsuit, ESG demanded that Section 40 of the Biological Diversity Act be declared "as unconstitutional and contrary to the principles enshrined in the Constitution" and that the notification be withdrawn (*ESG v. NBA*, 2012, 60). Indeed, ESG argued that Section 40 was arbitrary, as the criteria for determining what is a normally traded commodity had not been clearly elaborated, and that Section 40 was too broad in scope and therefore against the very spirit and object of the Act.

The public interest lawsuit prompted both NBA and KBB to take action. In November 2012, three years after ESG first raised the issue, NBA and KBB filed a criminal complaint before the Dharwad Bench of the High Court of Karnataka (*NBA v. UAS Dharwad*, 2012). In a curious development, court officers reported serving court summons to UAS Dharwad officials, but stated that they had been unable to locate company officials from either Mahyco or Sathguru to deliver the summons (Sood 2013).

In October 2013, Justice A. S. Pachhapure of the High Court of Karnataka ruled that, prima facie, there was sufficient material on record to determine that the developers of Bt brinjal had violated the Biological Diversity Act by failing to seek permission to access local eggplant varieties (*UAS Dharwad v. State of Karnataka*, 2013). On the normally traded commodities issue, the judge concurred that the exemption under Section 40 applied only to the export of a biological resource and not to research or industrial use. Finally, on the exception for collaborative research projects, Justice Pachhapure ruled that this exception was conditional on two criteria, neither of which had been met: that the project be approved by the Central Government, and that it comply with the policy guidelines. There was no evidence on file, for example, to show that a copy of the approval with all relevant documents had been sent to NBA, as required. This decision paved the way for reinstating criminal proceedings in the lower court in Dharwad. However, in 2014, the Supreme Court stayed the proceedings at the defendants' request. At the time of writing in 2021, the stay had not been vacated.

As for the public interest lawsuit, the High Court of Karnataka announced in December 2013 that the case dealt with environmental matters and would therefore be transferred to a National Green Tribunal (*ESG v. NBA*, 2013). The National Green Tribunals were set up in 2010 to handle cases pertaining to environmental issues, in an effort to relieve an overburdened court system and ensure the expedited resolution of environmental litigation by specialized judges.

In response, ESG filed a Special Leave petition with the Supreme Court of India in March 2014, challenging the decision by the High Court of Karnataka to transfer the case to a Green Tribunal. ESG argued that the case did not merely concern a violation of environmental law, but also challenged a provision of the Biological Diversity Act. The Green Tribunal, however, has no jurisdiction over the constitutional challenge to Section 40 of the Act; only a High Court or the Supreme Court can rule on such a challenge. As the issue of biopiracy was now the object of a criminal complaint, the challenge to Section 40 had become the vital component of the public interest lawsuit. As this book goes to press in 2022, the hearing regarding this special-leave petition in the Supreme Court has been postponed for seven years.

MONSANTO'S INTELLECTUAL PROPERTY RIGHTS TO BT BRINJAL

In India, Monsanto adopted a two-track approach to the commercialization of Bt brinjal: the private sector (Mahyco) was to focus on hybrid Bt brinjal varieties, whereas public agricultural universities were to pursue the development of open-pollinated varieties (NAS 2016).

The first issue of the ABSP-II South Asia newsletter published in September 2005 described the "momentous occasion" on which the seeds of Bt brinjal varieties that had been transformed in Mahyco's facility were handed over to the public agricultural universities:

Resource-limited farmers, burdened by yield losses in eggplant crops due to Fruit and Shoot Borer (FSB), now have a ray of hope. In a momentous occasion, Tamil Nadu Agricultural University Vice Chancellor Dr Ramasami received the backcrossed seeds of FSB-resistant eggplant from Dr. Usha B. Zehr, Joint Director (Research) of Maharashtra Hybrid Seed Company (Mahyco), one of the largest private hybrid seed companies in the country, on July 7, 2005 at the [Insect Resistance Management] meeting organized at [Tamil Nadu Agricultural University] in Coimbatore.

Subsequently, on the campus of the University of Agricultural Sciences, Dharwad's Vice Chancellor Dr. S. A. Patil received the backcrossed seeds of FSB-resistant eggplant from Mahyco on July 26, 2005. Aleen Mukherjee of USAID and Mr. Gopalakrishna of Sathguru Management Consultants were in attendance.

Vice Chancellor Dr. Ramasami remarked: "TNAU is committed to its mission of mitigating the worries of the resource-poor farmers. TNAU will deliver the transgenic FSB-resistant eggplant seeds to the farmers on low-profit basis. As a partner in the global consortium, TNAU identifies with the mission and vision of ABSP-II." (ABSP-II 2005, 6–7)

The ceremony reported above was the outcome of three separate agreements among Mahyco, Sathguru, and the agricultural universities.[16] Monsanto licensed rights to the eggplant event containing the Bt gene (known as the EE-1 event) to Mahyco. As Monsanto's sublicensor, Mahyco entered, in turn, into an agreement with each agricultural university for sublicensing the Bt gene for use in local eggplant varieties. According to these agreements, these Bt brinjal varieties then became "licensed domestic eggplant products" (Mahyco, Sathguru, and UAS Dharwad 2005).

Monsanto and Mahyco retained IP rights to these Bt brinjal varieties. The sublicense agreement states that "Monsanto/Mahyco IP Rights shall mean all IP rights that Monsanto or Mahyco owns or controls which will be infringed by making, using or selling Licensed Domestic Eggplant Products containing Mahyco Technology or Monsanto Technology (i.e., the Bt Gene)" (Mahyco, Sathguru, and UAS Dharwad 2005). In a statement made the context of the lawsuit, NBA (2013, 13) declared: "This seems to be an astonishingly wide IP claim over local varieties provided by UAS which Mahyco will backcross with their Bt brinjal, possibly violating Section 6 of the Biological Diversity Act."

The Agreement between Sathguru and Mahyco also has IP-related provisions. Under "Ownership," the Agreement states that "Mahyco retains all rights and title, including IP rights to the Mahyco material, Mahyco technology and Initial Bt Eggplant Products. Sathguru acknowledges and accepts that rights and title, including IP rights, to the Biological Materials including the Bt Gene vest in Monsanto Holdings Private Limited. Mahyco also retains all rights and title to all information, data, [and] records generated during the Project" (NBA 2013, 12).

In his moratorium decision, Environment Minister Jairam Ramesh noted that "the [Material Transfer Agreement] between TNAU and Monsanto in March 2005 has raised worrisome questions on ownership (both of products and germplasm) and what TNAU can do and cannot do" (MoEF 2010a). The nature of these worrisome questions is not made explicit in the decision. However, if one looks more closely at this Material Transfer Agreement, one notices that the only activity that TNAU can undertake is to adapt the semifinished cultivars provided by Mahyco to local growing conditions. The Material Transfer Agreement explicitly states that any other breeding activity is prohibited, and further specifies that Bt brinjal open-pollinated varieties can only be distributed "at cost." Finally, under no circumstances can the semifinished products provided by Mahyco be used by TNAU as parental lines for producing hybrids (Mahyco and TNAU 2005).

As discussed in the introduction, hybrid varieties are the products of a controlled breeding process called heterosis, or hybrid vigor. Contrary to open-pollinated varieties, whose seeds farmers can save and replant, the seeds of hybrid varieties cannot be saved without a significant reduction in both yield and quality. This creates an incentive for farmers to go back to the market and buy new seeds every year. For this reason, commercial seed companies mostly market hybrid varieties.

The Bt brinjal hybrid varieties and the open-pollinated varieties were not developed using the same germplasm. Mahyco's Bt brinjal (line 60208) uses a hybrid variety called RHR-51, bred by a major public agricultural university in the State of Maharashtra (NBA 2014). The open-pollinated Bt brinjal varieties, for their part, were developed using six local varieties from Karnataka and four from Tamil Nadu. It is these local varieties that are the object of the biopiracy complaint.[17] At least one of them—the Mattu Gulla variety cultivated in the village of Udupi—is well documented (see Appendix D).

In the public interest lawsuit, Mahyco argues that the ABSP-II project is not a commercial endeavor but rather is "pro-poor and not-for-profit" (Mahyco 2013). Mahyco contends that its role in the ABSP-II project "was limited to the donation of the technology to the public partners," that this donation was made free of cost or royalty, and that neither Mahyco nor its public partners has committed any clandestine acts with an intention to steal or misappropriate any biological resources covered by the Biological Diversity Act (Mahyco 2013). Sathguru is also emphatic that "there is no reverse flow of material or technology from UAS Dharwad to Mahyco," and that "Mahyco has not derived any commercialization rights to the varieties developed by UAS Dharwad" (KBB

2011, 147). NBA, however, firmly refutes this last claim in the criminal complaint:

It is pertinent to note that the Licensed Domestic Eggplant products claimed to be owned by Mahyco are in fact (a) Eggplant Planting Seed (as defined under Article 1.9 of the Agreement) produced out of (b) Eggplant Public Germplasm which are public-bred Eggplant varieties developed by public institutions (as defined under Article 1.11 of the Agreement) such as the UAS, and (c) this Eggplant Public Germplasm has been genetically modified by the UAS. From the above it is clear that not only Mahyco is guilty of genetically modifying local eggplant varieties made available by UAS without the approval of the Authority, but also Mahyco is guilty of laying proprietary claims to this modified eggplant seed. Such a proprietary claim is misconceived as the alleged Licensed Domestic Eggplant products use local biological resources as the raw material. (NBA 2013, 10–11)

Monsanto sought to obtain patent protection in India for the hybrid Bt brinjal that would be commercialized as part of the two-track approach if it were not for the national moratorium introduced in 2010. In 2006, Mahyco filed a patent application with the Indian Patent Office for a "transgenic brinjal (Solanum Melongena) comprising EE-1 event" (Mahyco 2006). According to the abstract:

The present invention relates to an insect resistant transgenic brinjal plant, plant cell, seed and progeny thereof comprising a specific event EE-1. Further, the invention provides the DNA sequence of the region flanking the insertion locus of the brinjal plant EE-1 event. It also relates to a process of detecting the presence or absence of specific brinjal plant EE-1 event. The invention also provides a diagnostic method for distinguishing the said specific brinjal plant EE-1 elite event in transgenic brinjal plants. The invention further provides a kit for identifying the transgenic plants comprising the elite event EE-1. (Mahyco 2006)

A First Examination Report was issued by the Patent Office in May 2013. According to this report, claims 9 through 11 fell under Section 3(j) of the Indian Patents Act 1970 on exclusions to patentability (CGPDTM 2013). The claims in question referred to transgenic plant or seed, plant cell, and progeny having the brinjal plant EE-1 elite event (CGPDTM 2013). Under Section 3(j) of the Patents Act ("What are not inventions"), microorganisms can be patented, but not seeds, varieties, species, or plants, either in whole or in part.

As happened with the patent application for both RR soybean and Bt cotton, Monsanto-Mahyco responded by deleting those claims that did not conform to the Patents Act and simply resubmitting the application. In response to the First Examination Report, Mahyco canceled four of the original claims made in the patent application (Lakshmikumaran and Sridharan 2014a).[18] In July 2014, Mahyco wrote to the Controller of Patents that, with the amendments made to the claims, the patent was now in compliance with the Patents Act, and requested that the patent be granted swiftly (Lakshmikumaran and Sridharan 2014b). However, following an intervention by NBA, the patent application for Bt brinjal was suspended until the legal dispute over the use of local eggplant varieties in the development of Bt brinjal is resolved (NBA 2014).[19]

So who owns Bt brinjal? There is no straightforward answer to this critical question. Monsanto holds patents on the event EE-1 in various jurisdictions, including the United States, but not in India. Moreover, the international public–private consortium that developed Bt brinjal used local eggplant varieties without the knowledge or permission of the local communities that grow these varieties, and without the authorization of biodiversity authorities, as required by the Biological Diversity Act.

The case raises fundamental issues regarding both the legal status of germplasm collection held by public agricultural universities and the rights of local communities.[20] In its report on the investigation, the KBB wrote that the UAS Dharwad had stated that, as an autonomous organization, it was exempt from the Biological Diversity Act. As KBB noted, however, there is no such exemption in the Biological Diversity Act (KBB 2010). This statement on the part of the university reflects a tacit understanding that public institutions such as UAS are exempt from the requirements of the Biological Diversity Act because they are public. As the Bt brinjal case illustrates, this assumption becomes problematic when public institutions enter into public–private partnerships.

As the Environment Minister noted in his decision on Bt brinjal: "doubts have been raised on how Bt-related research in these two institutions has been funded" (MoEF 2010a, 6–7). The dwindling of public funds for agricultural research and extension services since the 1990s means that hard-pressed public agricultural universities are looking for alternative sources of funding. It is an open secret that some public institutions have received funding as a result of their participation in ABSP-II. As their participation consists in providing access to local eggplant varieties, this raises uncomfortable questions about the way in which public–private partnerships can become conduits for the privatization of resources held by public institutions (Interview #15).

In its statement of objections to the public interest lawsuit, Mahyco argued that "There is no benefit-sharing involved, as the public partners involved, including the Indian Institute of Vegetable Research (IIVR), Varanasi, an ICAR institution; Tamil Nadu Agricultural University (TNAU), Coimbatore; and the University of Agricultural Sciences (UAS), Dharwad, plan to deliver the seeds to the farmers without any profit and therefore, 100 percent benefit goes to the farmers" (Mahyco 2013, 170–71). Underlying this statement is the assumption that farmers' communities actually want Bt brinjal and will benefit from it. The fact that farmers who grow the Udupi Mattu Gulla variety were not consulted as to the inclusion of this variety in the ABSP-II project, and opposed the project when they learned about it, shows that this is far from obvious (see Appendix D).

Mahyco also argued that "[The public partners] have not violated access to any biological material as the materials are their own breeding lines" (Mahyco 2013, 171). Plant germplasm held by public agricultural universities was collected from farmers' fields over the years with the understanding that it would be kept in the public trust. The statement by Mahyco, however, denies farmers any rights over this material.

Bt brinjal may be the first *national* case of biopiracy, in the sense that it is the first case filed before Indian courts. However, in its ramifications, the case is truly global. The Bt brinjal biopiracy lawsuit reveals the complex processes through which a US multinational company can claim IP rights over local varieties of eggplant held by state agricultural universities as a result of a bilateral deal on agriculture. In an era of privatization and public–private partnerships, the case also demonstrates how public institutions find themselves caught up in a tangled web in which notions of public interest and public good become highly equivocal.

5

PATENT POLITICS AND LEGAL ACTIVISM

Every seed makes a political statement.
—Manu Moudgil (2017)

For a decade after the commercial introduction of transgenic crops, confusion reigned about intellectual property (IP) rights over biotech seeds in Brazil and India. The introduction of transgenic crops in the early 2000s introduced wide-ranging changes to the legal landscape by requiring member countries to extend patents to microorganisms and microbiological processes, and to provide some form of property rights protection for plant varieties. The legislative changes enacted in Brazil and India to implement the World Trade Organization's Agreement on Trade-Related Aspects of Intellectual Property Rights (TRIPS) raised complex legal questions that were left unresolved. The main question concerned whether a biotech crop, as a plant variety, was protected under plant breeders' rights, or whether it was protected under patent law because it contained genetically engineered sequences. Some of the ambiguities in the wording of the TRIPS Agreement—for instance, regarding the definition of a microorganism under Article 27(3)b—were also transposed into domestic patent legislation. Since IP rights in agriculture had only recently been introduced in Brazil and India, expertise in this area was sorely lacking.

Finally, Monsanto's lack of transparency and deliberate obfuscation regarding its Brazilian and Indian patents added to the confusion.

The gray area was vast, and Monsanto exploited it fully. The fine line trod by financial traders—"We take advantage of the grey area between what we know we can't do and what we believe we can get away with"— could well describe Monsanto's approach to intellectual property (quoted in Brown 2018). Between 2002 and 2006, for example, the corporation charged an extremely high rate of royalties (or trait fees) on Bollgard I Bt cotton—equivalent to 75 percent of the total price of seeds. Remarkably, it did so despite the fact that it did not hold a patent in India. In Brazil, Monsanto continued to charge royalties on Roundup Ready soybeans for two and a half years after its patent expired.

The hype around transgenic crops in the early years meant that governments, scientists, and farmers in agricultural powerhouses like Brazil and India were anxious to gain access to the technology, and Monsanto ably exploited these sentiments. It resorted to lobbying and cooptation to impose unprecedented private systems of royalty collection. The model developed by Monsanto in the United States—consisting of strong patent rights, extensive licensing contracts signed by farmers upon the purchase of seeds, and an elaborate surveillance system to ensure compliance— was impossible to implement in countries like Brazil and India. In the latter country, for example, Monsanto was faced with millions of farmers on small land holdings, intractable legal enforcement issues, and the political impossibility of suing farmers. As Suman Sahai, chairperson of the Indian NGO Gene Campaign, observes: "Suing farmers for patent infringement would be committing suicide. You just don't sue farmers here in India" (Interview #9).[1]

As in the United States, the royalty collection systems implemented in Brazil and India were based on private contract law and sublicensing agreements. The modalities, however, were adapted to each country's crops and agrarian conditions. In Brazil, Monsanto implemented the collection of royalties on Roundup Ready soybean at the point where farmers sell their harvest to grain elevators and cooperatives. In India, Monsanto implemented the collection of royalties upstream in the cotton production chain, through the sublicensing agreements with the companies that produce Bt cotton seeds. In both cases, the royalty collection system effectively did away with the right to save seeds—by charging royalties on harvested grain in the case of Roundup Ready soybean, and by contractually restricting the Bt trait to hybrids in the case of Bt cotton. In short, while Brazil and India have provisions in their respective legislations allowing farmers to save seeds for replanting, the royalty collection systems implemented by Monsanto in effect rendered these provisions moot.[2]

While the Brazilian and Indian states may not have supported the introduction of proprietary rights regime as actively as their United States and Canadian counterparts (Pechlaner 2012), they have certainly been complicit in their implementation. I can only concur with Felipe Filomeno's assessment that in South America, "the implementation of Monsanto's system of royalty collection relied on coercion and cooptation of some associations of rural producers, local seed companies and national governments, bringing its legitimacy into question" (2014, 13–4).

Shalini Randeria's conceptualization of the constrained agency of subordinate states in the Global South allows for a more nuanced understanding of the role of the state in the implementation of IP regimes for biotech crops (2003a, 2007). According to Randeria, states in the Global South remain pivotal in selectively transposing neoliberal policies and international norms to the national terrain, while at the same time capitalizing on their perceived weakness in order to render themselves unaccountable to their citizens (Randeria 2003b). Hence, in the case of Bt cotton, the Indian government hid behind its international obligation to implement Article 27.3(b) of the TRIPS Agreement to grant exclusive rights to Monsanto when the latter, in fact, did not hold a valid Indian patent. As Carl Pray and Latha Nagarajan (2010, 300) observe in the Bt cotton case, the "Indian regulatory system gave [Mahyco Monsanto Biotech[3]] a temporary monopoly on the Bt gene."

Instead, both the Brazilian and the Indian governments could have used the flexibilities available in the TRIPS Agreement to set limits to IP rights on biotech crops in the interest of farmers—especially since Article 27(3)b was still under review. As Philippe Cullet (2005b) observes, India could have introduced restrictions on the patentability of microorganisms in conformity with its own TRIPS-compliant Patents Act. These restrictions could have taken the form of a provision stating that "micro-organisms are only protected in isolation and not where they are inserted into another organism which is itself not patentable under the Patents Act" (Cullet 2005b, 3609). This interpretation would have been consistent with the exclusion of seeds in Article 3(j) of the Indian Patents Act and would have prevented the courts from interpreting a patent owner's right over a genetic sequence as extending to seeds and plants, as in the case of *Monsanto v. Schmeiser*. Governments could also have established that patent rights and private sublicensing agreements could not override farmers' rights to save seeds under plant variety protection law. The following account by Peter Newell is indicative of the influence of corporate lobbying and of governments' omission at a critical juncture in the implementation of IP regimes for biotech crops:

When Argentina called a meeting of Ministers of Agriculture in Mercosur in 2005 to generate support for its position against paying Monsanto royalties on soya crops (rather than seeds), initial support was forthcoming from Brazil and Paraguay. Intense pressure in the wake of the meeting, however, led to these governments retracting their positions on the basis that they were concluding their own agreements between the private sector and Monsanto. (Newell 2008, 263)

In sum, various options were available to regulate IP rights on biotech seeds under the TRIPS regime, but these were not explored, and thus Monsanto was given a free hand to proceed as it pleased.

LEGAL ACTIVISM AROUND INTELLECTUAL PROPERTY AND BIOTECH SEEDS

As farmers and other actors in the soybean and cotton production chain felt the impact of these new proprietary rights systems, they began to question Monsanto's IP rights and practices in national courts. Farmers were put off by what they perceived as the excessively high cost of royalties and seeds, and they resented Monsanto's aggressive and intrusive practices, such as performing GMO detection tests on farmers' grain when they showed up at the grain elevator. This sentiment was expressed by one grain cooperative manager in the Brazilian State of Rio Grande do Sul, who told his member of Congress, "I can't stand it anymore—seeing those Monsanto people showing up at the grain elevator and behaving as if they own everything."[4] Luiz Fernando Benincá, the Brazilian soybean producer, did not mince his words: "As [Monsanto] is amoral, it will do anything for profits. It does not respect anything. It ends up committing the worst crimes against nature and against people. Whoever gets in its way gets eliminated" (Interview #33).[5] One might expect to hear this statement from a member of one of Brazil's left-wing agrarian movements. These words, however, were spoken by a politically conservative large landowner and illustrate my contention that legal disputes around intellectual property and biotech crops have brought together strange bedfellows.

Indeed, those challenging the legitimacy of IP rights over biotech seeds are not those who are involved in litigation focusing on health and environmental regulations, and who are part of the broader movement for food and seed sovereignty. As described in chapter 2, the Roundup Ready soybean class action in Brazil originated when Luiz Fernando Benincá, a large soybean farmer who felt deeply dissatisfied with the royalty collection system, approached his lawyer after failing to obtain the backing of his own federation. Feeling his federation had been coopted, he filed a class action lawsuit through his local rural union. Two more rural unions and, importantly, the state federation of family farmers (FETAG-RS) joined the class action shortly afterward. In the polarized Brazilian countryside, the class action thus represented a rare example of an alliance between large rural producers and small farmers. If differences in land access and ownership are key differentiating factors among rural-based working classes and groups (Edelman and Borras 2016), all farmers rely to some extent on access to seeds and the ability to reproduce them.

In the case of Bt cotton in India's State of Andhra Pradesh, two left-leaning farmers' organizations affiliated with the Communist Party of India—the All India Peasant Union and the Andhra Pradesh Rythu Sangam—filed an initial complaint against Monsanto in 2005.[6] In the more recent phase of the dispute before the Delhi High Court (2015 onward), national as well as multinational corporations—including Indian seed companies and Monsanto—have replaced farmers' organizations as the main protagonists. Domestic seed companies have enlisted the support of influential Hindu nationalist organizations, including those representing farmers, against Monsanto.

Finally, the Bt brinjal biopiracy case was spearheaded single-handedly by the small Indian NGO Environment Support Group, which uses strategic litigation to advance environmental causes, while the larger GM-Free India coalition did not actively support the case. As can be seen from this brief overview, those who have engaged in litigation around IP rights to biotech crops are remarkably diverse. They are also distinct from the actors that have mobilized against GM crops more broadly—the GM-Free Brazil Campaign and the Coalition for a GM-Free India.

Why then, one wonders, have food and seed sovereignty activists not played a greater role in legal challenges over intellectual property and biotech crops? This is certainly not due to a lack of concern. In fact, activists have long been critical of biotech patents and royalties, and have been among the first to raise concerns regarding their impacts on agro-biodiversity as well as farmers' livelihoods. However, the high cost and prolonged nature of litigation represent important barriers for organizations with limited financial and human resources. Asked in May 2016 whether she believed Monsanto's patents were in compliance with the Indian legislation, one prominent Indian activist responded, "To be honest, no one really knows. The problem is that these patents have never been tested in the courts." But, she went on to deplore, "who among us can afford to take on such a legal challenge?" (Interview #9).

With the exception of the Indian seed companies in the Bt cotton case, which as commercial actors have more resources than civil society actors, the two sides have vastly unequal means and resources. Monsanto employs a huge professional legal team and sets aside vast sums of money to cover potential litigation costs.[7] Farmers' unions and NGOs, in contrast, have extremely limited financial resources and typically rely on pro bono lawyers. In both the Roundup Ready soybean lawsuit and Bt brinjal public interest litigation, lawyers representing farmers have worked on a pro bono basis, at a great personal cost, for over a decade. Asked why he had embarked in this class action lawsuit, the lawyer acting on behalf of farmers' unions told me, "I have dedicated a significant part of my professional life to this class action. Why? Because I think this is right. For no other reason. I think that what they are doing is wrong. And one day—maybe not now, but one day—we will win" (Interview #29A).

The fact that litigation is costly and time consuming partly explains why food sovereignty activists with limited resources may have been reluctant to initiate legal proceedings. But why, then, did they fail to actively support these lawsuits when they were filed by other actors? I believe the reasons are both strategic and ideological, and reflect a keen awareness of the risks of cooptation. First, some activists are critical of litigation as a strategy to achieve their goals. Commenting on the Bt brinjal case, one food and seed sovereignty activist opined that the legal route can, under certain circumstances, represent a good short-term tactic in order to buy time. Hence, public interest litigation may be warranted to prevent the impending commercialization of new GM varieties, as has been done for fifteen years by public interest lawsuit no. 260/2005 (*Aruna Rodrigues v. Union of India*, 2005). As of 2020, this lawsuit had prevented the environmental release of Bt brinjal and GM mustard in India, in the absence of a proper regulatory process and biosafety protocol. According to the same activist, however, it is not necessarily a good long-term strategy, because it relies too heavily on receptive officials within regulatory institutions; when these individuals leave or are transferred, litigants are back to square one. This is what that activist thought had happened with the Bt brinjal biopiracy case. She did not have much faith in the effectiveness of protecting biodiversity and farmers' interests by enforcing the existing legislation. She believed, for example, that the individualistic approach of the Indian Protection of Plant Varieties and Farmers' Rights Act was merely pitting farmers against one another and that, rather than preventing biopiracy, the Biological Diversity Act could, in practice, allow corporations smoother access to traditional knowledge (Interview #14A). These views on the limits of legal mobilizations and of the implementation of existing legislation echo the sentiment of many food sovereignty activists elsewhere. As one NGO member from Colombia put it, "We have grown tired of legal activism. Even when we win, the state manages to turn things in their favor" (quoted in Silva 2017, 155).

Activists also perceive the protracted nature of legal battles as an important drawback. Temporality plays out differently, depending on the nature of the issue, and the fact that seeds are living entities has important implications. As Shalini Randeria and Ciara Grunder (2011) argue in their study of evictions in urban India, litigation can be used strategically by poor city dwellers to "stretch time" and delay forced displacement. However, in the case of seeds, prolonged court cases mean that transgenic varieties have the opportunity to spread, whether legally or illegally. As living entities that reproduce, seeds have the ability, so to speak, to evade formal legal processes. In Brazil and India, the widespread cultivation of unapproved GM varieties in the early 2000s put pressure on governments to authorize their cultivation (Herring 2007), and history repeated itself with herbicide-tolerant (HTBT) cotton in India in the late 2010s (Jadhav 2019). In this context, some activists feel that they cannot afford long-winded legal processes that face an uncertain outcome. In addition, until a dispute is settled, a litigant company can continue to charge royalties on a genetically modified product, so delays in court benefit the corporation, as in the case of Roundup Ready soybean in Brazil.

Moreover, many activists express a healthy skepticism regarding IP rights. Proprietary issues surrounding transgenic crops are often deemed less urgent than is preventing their environmental release. More important, activists are concerned that engaging on IP-related issues might ultimately contribute to legitimizing existing IP regimes for plant varieties. For example, when the Indian government announced in 2016 that it would hold public consultations on the Licensing Guidelines for GM technology agreements, GMO critics were faced with a dilemma: should they participate in the consultations in order to influence the outcome and perhaps arrive at a royalty collection system that would be fairer for farmers? Or should they avoid the consultations, since doing so would amount to legitimizing the royalty collection system? (Interview #14B). This is a classic example of the risks of cooptation that social movements face when they engage in institutional processes. By participating in processes dominated by influential actors, they fear that they may end up endorsing the status quo rather than challenging it.

Indian seed sovereignty activists, for example, see the government regulation of Bt cotton as a double-edged sword: while they welcome efforts to curb corporate practices considered to be predatory, they also feel wary that capping Bt cotton seed prices and royalties could make Bt cotton cultivation more attractive to farmers. This illustrates Michael McCann's argument about the twofold potential of legal mobilization that offers both opportunity and constraint, in effect generally upholding the status quo but at times providing limited opportunities for challenges and change (McCann 2004). This concern about the double-edged dimension of activism around intellectual property and biotech crops was echoed by one European activist: in mounting legal challenges to IP rights based on biopiracy, the coalition of which he was a part deliberately steered clear of biotech patents, privileging litigation around non-GM plant resources instead (Interview #71). In the context of the legal disputes concerning Roundup Ready soybean, Bt brinjal, and Bt cotton, activists at the same time felt highly critical of Monsanto's IP practices and were keenly aware of the risks of cooptation involved in lending their support to disputes that were driven in large part by the short-term commercial interests of other actors in the agbiotech economy, such as large soybean growers and seed companies.

The dilemmas that food sovereignty activists face took on an additional dimension in the context of the Bt trait-fee controversy. The ideology of food sovereignty activists, rooted in transnational solidarity among peasant organizations, stands at odds with the ultranationalist ideology of right-wing Hindu organizations, and yet the two converge in their critique of GM crops. Food and seed sovereignty activists oppose GM crops because they consider these crops to be detrimental to the environment and to biodiversity, besides promoting corporate concentration at the expense of farmers. Hindu nationalists, for their part, emphasize that these technologies owned by multinational corporations are not *swadeshi*—literally, "of one's own country"—and therefore have integrated the idiom of seed sovereignty in their discourse. In the words of one leader of the farmers' organization BKS (Bharatiya Kisan Sangh), part of the Hindu ultranationalist RSS movement, "Monsanto should go back, as it is important for seed sovereignty. We can produce our own seeds like we did in the past" (Agha 2018; see also Bhardwaj, Jain, and Lasseter 2017). The BKS has also labeled Monsanto a "threat to seed sovereignty" (Andersen and Damle 2019, 142).

Some RSS critiques of GM crops are couched in an ultranationalist and essentialist discourse about "natural food" and the inherent value of desi plant varieties. The term *desi* refers, in this context, to plant varieties that are thought to be native to the Indian subcontinent. The line between discourses of seed sovereignty and ultranationalist discourses is sometimes blurred: environmental activist Vandana Shiva, for example, uses nationalist and essentialist tropes when she writes about the "clash of civilization"[8] between India and the West and about "India's ancient love for nature" (Shiva 2016b).[9]

Despite their political differences, many left-wing and right-wing activists throughout India share a common concern about corporate concentration of power and the erosion of agrobiodiversity. Hence, organizations from both sides of the political spectrum—for example, Greenpeace and RSS organizations—have lent their support to the "Monsanto Quit India" Campaign and have opposed GM crops (Mohan 2016). These ironies are not lost on the activists themselves. A leader of the RSS economic wing observed to me, with an amused smile, that he had found himself speaking at a press conference on GM mustard organized by civil society and sharing the podium with a number of well-known food sovereignty activists who otherwise would not have stood on the same platform with him because they espouse radically different political views (Interview #68).

More fundamentally, food and seed sovereignty activists' reluctance to engage in legal mobilizations on IP issues stems from the evolution of their views and analyses concerning intellectual property and the legal status of seeds. In the course of the past three decades, a shift within the food and seed sovereignty movement has been observed toward pursuing strategies outside of formal legal frameworks (Peschard and Randeria 2020). At the end of the day, some activists hold the conviction, born out of the arduous and ultimately failed efforts to secure farmers' rights through international negotiations and agreements, that the sole effective way to protect farmers' varieties is to work at the grassroots level to keep seeds in farmers' hands, instead of relying on the courts. The outcome of the Bt trait-fee dispute—an out-of-court settlement putting an end to litigation and preempting a court ruling—suggests that they are right to be cautious.

The motivations of litigants in these lawsuits were diverse: a deep sense of dissatisfaction with corporate practices, strong nationalist sentiments, competing business interests, concern about the biopiracy of farmers' plant varieties and traditional knowledge, and a commitment to farmers' rights. Litigants did not set out to mount an outright challenge to the corporate food regime. And yet in some cases, the processes they set in motion reached beyond what some of the parties to the disputes had initially envisioned. In the case of Bt cotton, for example, Indian seed companies sought government regulation of the royalties they paid to Monsanto. The government then intervened to regulate not only royalties but also seed prices, a decision that benefited farmers but not seed companies, leading an observer to comment that seed companies had shot themselves in the foot by seeking government intervention (Fernandes 2018). More important, lawsuits initially concerning royalties evolved to encompass broader questions regarding patents. The irony is that these lawsuits ended up achieving what some seed and food sovereignty activists had been longing for: questioning the very validity of the biotech patents at the heart of the new proprietary-rights regime in agriculture.

TOWARD A NEW LEGAL COMMON SENSE?

The judiciary plays a key role in validating patents, especially when intellectual property expands into uncharted waters, as was the case with the extension of patents to microorganisms and microbiological processes in the mid-1980s. As a Monsanto Canada spokesperson declared in 2001 during the patent infringement lawsuit against Canadian farmer Percy Schmeiser, "we did have a number of people waiting in the queue, but [Schmeiser] was the first case where we attempted to find out if the patent was valid. You don't know how strong that patent really is until somebody violates it and it's upheld in a court of law" (*Canadian Press* 2001). In both Brazil and India, one crucial reason Monsanto has enjoyed a free rein to implement private royalty collection systems was that the legality of its biotech patents under the domestic legislation had not yet been tested in the courts.

Until the 2018 decision of the Delhi High Court in *Nuziveedu v. Monsanto*, the only case concerning the patentability of living organisms in India was a 2002 judgment of the Calcutta High Court in *Dimminaco AG v. Controller of Patents*. The Indian Patent Office had rejected an application by the Swiss biotechnology company Dimminaco AG for a patent on a method for producing a live vaccine, on the grounds that a process resulting in a living substance was not patentable under the Patents Act. Dimminaco appealed and the Calcutta High Court overturned the decision, stating that "there is no statutory bar to accept a manner of manufacture as patentable even if the end product contains a living organism" (*Dimminaco AG v. Controller of Patents*, 2002). For the purposes of this discussion, it should be noted that the case was limited to the patenting of a *process* and not of a *product*, that the case did not involve a higher life form, and that the microorganism in question was not transgenic. As for Brazil, there had been no litigation involving patents on living organisms before the legal challenges discussed here.

In her comparative study of controversies around patents on life forms in Europe and the United States, Shobita Parthasarathy (2017, 156) emphasizes "the long-standing differences in how the two jurisdictions saw patents—as techno-legal in the United States, and as moral and policy objects in Europe." As she observes, "US patent-system institutions were reluctant—and often simply lacked the capacity" to explicitly consider moral and socioeconomic concerns, not to mention the distributive implications of patents (Parthasarathy 2017, 174). Hence, when the Organic Seed Growers and Trade Association filed a lawsuit in the United States arguing that Monsanto's enforcement of its patent rights was harming both farmers and consumers, the Court dismissed the lawsuit as baseless and "derided the plaintiffs' efforts to use the court to address moral and regulatory issues" (Parthasarathy 2017, 173).

In this regard, Brazil and India are closer to the European patent culture than to that of the United States. In the post-independence period, both countries have pursued policies that sought to balance IP rights with industrialization and the public interest. Their diplomatic turnaround in the WTO negotiations did not transform the broader patent cultures overnight. But in the absence of clear policies and case law, their respective patent offices were left to their own devices to interpret the new legislation and then to make complex decisions about the first patent applications involving transgenic plants.

The Indian Patent Office publishes and regularly updates a *Manual of Patent Practice and Procedure*, to provide guidance to patent examiners. The 2005 edition explicitly stated that genes were not considered patentable (Ravi 2013, 324). For undisclosed reasons, that statement was deleted from the 2008 edition of the manual (OCGPDT 2008). The *Guidelines for Examination of Biotechnology Applications for Patent*, published in 2013, state that patents can be claimed on, inter alia, polynucleotides or gene sequences, polypeptides or protein sequences, gene constructs or cassettes, microorganisms, transgenic cells, and plant tissue culture (OCGPDT 2013, 4). The mere discovery of any living thing occurring in nature is not considered a patentable invention (OCGPDT 2013, 11).

The first patents on biotechnology were granted in India following the last amendments to the Patents Act in 2005 (GoI 1970). That year, 73 patents were granted. The pace picked up in the following years, with 1950 applications and 314 grants in 2007–2008, the last year for which data is available (Singh 2015, 108). Under the heading "Patent grants by the [Intellectual Property Office]: Is there a method in the madness?," Bhavishyavani Ravi (2013) seeks to identify the criteria used by the Indian Patent Office to assess patent applications involving nucleotide sequences. The heading is a reference to the apparent inconsistencies in how patent applications related to genetic material were treated by Indian patent offices. However, according to Ravi, there was a consensus among the patent examiners he interviewed that the exclusion referring to plants and plant parts in the Patents Act were not applicable at the molecular and cellular level when genes were involved (Ravi 2013).[10]

Chan Park and Arjun Jayadev (2011) observed, with reference to pharmaceutical patents, that the dearth of Indian patent case law since the amendments to the 1970 Patents Act has meant that courts and patent offices have relied on foreign judgments to interpret the basic criteria for patentability. Stated differently, the fact that these traits had obtained patent protection in major patent jurisdictions such as the United States and Europe weighed in considerably when patent offices in other countries such as Brazil and India examined these patent applications, and overrode the fact that these countries had different patent laws. This can explain the fact that the Brazilian Patent Office made a submission to the court in support of revoking the patent it had granted to Monsanto on Intacta soybeans in 2012 (Fincher 2012). The Patent Office declared that it had erred in granting patent PI 0016460–7, since the latter combined already-existing technologies (the Roundup Ready and Bt traits) and therefore did not meet the inventive-step criterion (Tosi 2018). While the Patent Office did not explain why it had initially granted the patent, one can infer that its decision at the time might have been influenced by the fact that Intacta RR2 PRO had been patented in the United States and other jurisdictions. Patents, however, are granted at the national or, at most, regional level, thus foreign judgments on patents are not legally binding in other jurisdictions (Park and Jayadev 2011). Therefore, the possibility remains that Brazilian as well as Indian courts could forge their own jurisprudence in evaluating the basic criteria for patentability and in balancing IP rights with other concerns, such as farmers' rights and food security.

There are signs that this is starting to happen with regard to agbiotech patents. Notably, the April 2018 ruling by the Delhi High Court in the patent infringement case opposing Nuziveedu to Monsanto began to make up for the lack of case law and offered the chance for the first interpretations by the Indian judiciary on the patentability of biotech seeds. Overall, however, these two trends—countries relying on foreign jurisprudence versus forging their own jurisprudence—compete with each other in the decisions delivered by Brazilian and Indian courts so far.

In the three case studies discussed in previous chapters, the main judicial decisions in favor of Monsanto are narrowly grounded in patent law. In the Roundup Ready soybean class action, Brazil's Superior Court of Justice ruled that as a product of genetic engineering, biotech seeds come under the exclusive protection of the Industrial Property Act, and that those who opt for them must compensate the patent holder for the use of the technology. The Court accepted the argument that exclusive rights granted to a patent owner can extend to a cultivar and dismissed the Plant Variety Protection Act as altogether irrelevant to the case (*Sindicato rural de Passo Fundo v. Monsanto*, 2019).

This line of interpretation mirrors those of the Supreme Courts of the United States and Canada in major rulings on intellectual property and GM crops (for example, *Monsanto v. Schmeiser* 2004; *Bowman v. Monsanto*, 2013) and often explicitly builds on US and Canadian case law. For example, in its 2017 decision on Bt cotton, India's Delhi High Court argued that the reasoning of the Supreme Court of Canada in *Monsanto v. Schmeiser* was "weighty, deserving to be adopted by this court" (*Monsanto v. Nuziveedu*, 2017, 80). More specifically, Judge Gauba adopted the Canadian

Supreme Court interpretation that the fact that a patented object or process was part of a broader unpatented structure (that is to say, a plant) was ultimately irrelevant. Based on this reasoning, the Delhi High Court ruled that Indian seed companies' activities—generating hybrid varieties of cotton seeds through biological processes—could be construed as infringing on Monsanto's patent, even if essentially biological processes for the production or propagation of plants are excluded from patentability under the Indian Patents Act (*Monsanto v. Nuziveedu*, 2017). Similarly, in the Roundup Ready soybean class action, the Brazilian court of appeal referred to a key argument used by the US Supreme Court in *Bowman v. Monsanto* (2013). The argument concerned a tricky issue created by the extension of patent rights to life forms: when do the exclusive rights of a patent owner end in cases where the invention is a self-replicating organism (or become "exhausted," in patent parlance)? In *Bowman v. Monsanto*, the US Supreme Court ruled that patent owners' rights extended to successive generations of the plant. This reasoning was adopted by Brazil's Court of Justice of Rio Grande do Sul in its ruling that overturned the lower court decision by Judge Conti (*Monsanto v. Sindicato rural de Passo Fundo*, 2014). In short, there is no bar to relying on foreign decisions. In these cases, however, that reliance meant glossing over the significant legal differences between, on the one hand, the Brazilian and Indian legislation and, on the other, those of the United States and, to a lesser extent, Canada.

In contrast, the main decisions against Monsanto strove to interpret the issue in light of a wider set of legal norms, including each country's respective constitutions and domestic laws related to patents and plant variety protection.

In his first decision in the Roundup Ready soybean class action, for example, Judge Conti took a more restricted view of a patent holder's rights, and argued that the Plant Variety Protection Act (PVP Act) should take precedence over the Industrial Property Act when it comes to plant varieties. To support this interpretation, he pointed to the fact that the PVP Act (1997) was passed one year after the Industrial Property Act (1996), as "the sole form of protection in the Country for plant varieties" (Art. 2), thus reflecting an intent to submit plant varieties to a distinct legal regime. This intention, he added, was further exemplified by Brazil's decision to join the 1978 Act—and not the more restrictive 1991 Act—of the UPOV Convention (*Sindicato rural de Passo Fundo v. Monsanto*, 2012, 14).

The dissenting opinion, presented by Judge Lopes do Canto, in the class action suit was informed by broader concerns over food security and over the limits and social function of property rights. According to this dissent, the Brazilian Constitution holds that "No property right is absolute and can prevail over its most relevant social functions" (*Monsanto v. Sindicato rural de Passo Fundo*, 2014, 65). Judge Lopes thus redefined the conflict as one between a third party's intellectual property and the rights guaranteed to small farmers in the Constitution.

According to that judge, no perpetual rights inhere in plant breeding itself (*Monsanto v. Sindicato rural de Passo Fundo*, 2014, 73). Judge Lopes reasoned that Monsanto holds property rights over the initial technology, but that these do not extend either to the entire production process or to successive generations of plants. In his opinion, charging royalties on harvest represented an attempt to obtain financial gains far superior to an equitable remuneration for use of the technology. He argued that the patent holder can charge royalties on the sale of seeds to farmers, but insisted that patent rights are exhausted from then onward. In other words, he took the view that patent law is no longer applicable when a farmer sells his harvest as food or raw material, sets aside and replants seeds, or multiplies seeds to give or exchange, or if the cultivar is used for plant breeding or scientific research (*Monsanto v. Sindicato rural de Passo Fundo*, 2014, 67).

In sum, Judge Lopes do Canto argued that since a specific law exists, having been passed with the objective of protecting the country's small farmers in compliance with the Constitution, this statute must prevail if a conflict is seen with another law: "When there is a normative conflict, the social interest must prevail over purely private interests. In other words, the law that must be applied is the one that best serves collective interests, in this case, the PVP Act" (*Monsanto v. Sindicato rural de Passo Fundo*, 2014, 73). Given the importance of family agriculture for Brazilian food security, he concluded, it was essential to guarantee the right to plant freely in the interest of society.

Indian courts have raised a similar set of considerations in some of their rulings. In the 2017 ruling of the Delhi High Court, the judge observed that Article 39 of the Constitution of India mandates the State to "direct its policy towards securing that the ownership and control of the material resources of the community are so distributed as best to subserve the common good" (*Monsanto v. Nuziveedu*, 2017, 35). In the same decision, the judge also noted that the Essential Commodities Act was enacted in the public interest (*Monsanto v. Nuziveedu*, 2017, 37).

As for the 2018 ruling of the Delhi High Court, it was remarkable for being the first to examine the legality of patents on agbiotech traits under Indian law and, further, to delve into the substantive issues. The ruling stood out for a number of reasons. First, the judges addressed the fact that the claims made in the patent application had to be modified substantially to conform to the national legislation. On account of Section 3(j) of the Indian Patents Act, which covers exclusions to patentability, the Patent Office had rejected more than half the claims made in the original patent application. These claims were related to plants, plant cells, tissues, and progeny plants containing the nucleic acid sequence, as well as plants created through an essentially biological process. Of the remaining 27 claims that were granted, 24 were process claims, and only three were product claims related to a nucleic acid sequence. According to the judges, this narrowing of the patent claims was relevant and had implications for the scope of protection granted by the patent (*Nuziveedu v. Monsanto*, 2018).

Second, the judges interpreted patent rights over biotechnological inventions in light of India's distinct legislation in the area of agricultural patents and farmers' rights. In his decision, the single-bench judge had relied on *Monsanto v. Schmeiser*. The division bench judges rejected this line of reasoning, arguing that the uniqueness of the Indian legislation sets it apart from the United States and Canada and that *Monsanto v. Schmeiser* could therefore not be extrapolated to India. The judges also noted that the Indian Protection of Plant Varieties and Farmers' Rights Act (PPVFR Act) grants substantive rights to farmers, in contrast to the United States and Canada, which do not formally recognize farmers' rights.

Third, the judges questioned the interpretation that biotech traits are transgenic microorganisms. They rejected Monsanto's claim that the subject matter of the patent is a microorganism, patentable under TRIPS Article 27.3(b). They argued, instead, that a nucleic acid sequence is not a microscopic organism, because it has no existence of its own. It is only of use after it is introgressed into seed material, which must in turn undergo hybridization. The judges recognized that Monsanto can assert patent rights over the nucleotide sequence responsible for the Bt trait. However, they argued that the trait has no intrinsic worth. It only becomes valuable if it is part of a plant cell or seed, both of which are explicitly excluded from patentability under Section 3(j) of India's Patents Act.

Fourth, the judges argued that the transfer of the Bt trait to plant varieties through hybridization is an essentially biological process, which is also exempted from patentability under Section 3(j). Under the sublicensing agreement, Monsanto supplies donor seeds incorporating the Bt trait to a seed company, which then uses the donor seeds to transfer the Bt trait to its own varieties through conventional breeding techniques. The judges concluded that the moment the DNA containing the nucleotide sequence (the subject matter of the patent) was hybridized to produce the transgenic seeds or plants, the latter fell within the purview of the PPVFR Act, the Indian legislation regulating plant breeders' rights.

These decisions and dissenting opinions show that there are no foregone conclusions when it comes to the patenting of genes, seeds, and plants. On the contrary, the extension of intellectual property to plant materials left many unresolved issues and much room for alternative legal interpretations grounded in a country's domestic legislation, patent culture, and political priorities. By remanding the case to the Delhi High Court, the Supreme Court of India missed the opportunity to rule on a matter of considerable public interest. The Supreme Court could have argued, as it did in *Novartis AG v. Union of India*,[11] that the matter warranted an expeditious decision. In any case, the out-of-court settlement reached by the parties meant that there would be no ruling on the patentability of genes for the time being.

I would like to end this chapter by briefly discussing a legal ramification of the dispute over Monsanto's Roundup Ready patent. There is no question that the Brazilian patent on Roundup Ready (RR1) soybean expired in August 2010. However, as I detailed in chapter 2, Monsanto sought to obtain an extension of the term of protection of its patent in Brazil in line with the extended term it had obtained in the United States, and then sued the Brazilian Patent Office when it denied the extension. After having lost in three instances, Monsanto filed an extraordinary appeal to the Supreme Court of Brazil. The rules concerning the term of protection for pipeline patents are laid out clearly in the Industrial Property Act, and Monsanto was unlikely to win the case. Yet by using all the appeal mechanisms available, Monsanto was able to delay for years a final decision on the validity of its patent on Roundup Ready (RR1) soybean.

In the meantime, the judgment of the extraordinary appeal to the Supreme Court was suspended due to a Direct Action of Unconstitutionality (ADI) concerning the pipeline provisions of the Industrial Property Act (ADI 4234 DF 2009). An ADI is a Brazilian legal instrument that allows a challenge in the Federal Supreme Court to any law whose constitutionality may be in question.[12] In this case, the Attorney General of the Union[13] argued that pipeline patents were unconstitutional because they allowed, to the detriment of the novelty principle, the patenting of something that was already in the public domain, thus fostering the expropriation of the common good. As I laid out in chapter 1, pipeline patents—also called revalidation patents—allowed the retroactive patent protection of inventions that were in the public domain in Brazil but were patented abroad, without requiring a technical examination of the patentability requirements.

The Federal Supreme Court received the ADI in April 2009. The president of the Court determined at the time that the case would follow an expedited process, in which an ADI is sent directly to the full Supreme Court for final decision. However, in a decision entirely at odds with the spirit of the expedited process, the Court president then took nine years to schedule the case for judgment, and the case was again removed from the agenda in June 2021. As ADI 4234 was proposed more than 10 years ago, and since there are no more pipeline patents in effect, the Court may decide not to rule on its merits on the grounds that the case has lost its object. If this happens, it will confirm the view that the court chose to let the case die, in a case of "deliberate omission" on a matter of high relevance for both public health and agriculture. As Soraya Lunardi and Dimitri Dimoulis (2017) argue, this inaction had serious consequences. Pipeline patents granted in 1997 have expired and the inventions they cover thus entered the public domain in 2017 at the latest. The inaction of the Supreme Court meant that Brazilian society paid the significantly higher cost of patented medicines and patented seeds during those years (2009–2017) without the Supreme Court having ruled on the merit of the ADI.

In the case of Roundup Ready soybeans, the combined effect of the lawsuits against the Brazilian Patent Office and of the delayed judgment of the ADI was to prolong the uncertainty surrounding the Roundup Ready soybean patent. The ruling of the Superior Court of Justice in the class action, delivered in October 2019, stated that owing to the extraordinary appeal and ADI itself, the term of protection of the Roundup Ready (RR1) soybean patent was still open to question (*Sindicato rural de Passo Fundo v. Monsanto*, 2019, 17).

Litigants who set out to challenge IP rights on biotech seeds in Brazil and India have had mixed success in the courts. As Boaventura de Souza Santos (2002) reminds us, although the law *can* be emancipatory, it is never inherently so. The significance of these legal challenges lies elsewhere: by drawing attention to the role of both corporations and governments in the implementation of royalty collection systems for biotech crops, litigation has revealed the political nature of biotech patents as well as royalty collection systems. These legal challenges have destabilized the dominant pro-biotech interpretations of patent rights that had been endorsed by the Supreme Courts of the US and Canada. Such legal challenges have shown that within the minimum norms established by the TRIPS Agreement, considerable leeway still exists to balance patent rights against other considerations, such as food security and farmers' rights. Most important, these lawsuits have forced the courts to begin to examine the complex issues raised by the extension of intellectual property to plant-related material in the context of each country's constitutions, domestic laws, and policy goals. Amid growing concern over historically unprecedented levels of concentration in the global seed industry, these lawsuits offer insight into the emergence of a new legal common sense concerning the merits and limits of extending intellectual property to plants and seeds.

CONCLUSION

As this book goes to press, all three lawsuits have reached their countries' highest courts. In the Bt cotton trait fee dispute, the out-of-court settlement between Bayer-Monsanto and Nuziveedu has put an end to all ongoing litigation. In the Bt brinjal case, both the public interest litigation and the criminal prosecution are before the Indian Supreme Court but have been in limbo for years. As for the RR soybean class action, all the appeals have been exhausted, though the farmers' unions and their lawyers do not rule out bringing a rescissory action questioning the Court's understanding that the object of the patent is a microorganism.[1] The rural unions are also considering filing a nullity action against 18 biotechnology patents that have already been granted as well as a number of patent applications under examination.

Independently of the final outcome, the decisions rendered thus far have already broken with the dominant paradigm by offering legal interpretations that balance the rights of patent holders against those of farmers, food security, and the public interest. Indeed, these lawsuits have evolved to challenge fundamental dimensions of the corporate food regime in agriculture.

First, these legal challenges aimed to redefine the relationship between public and private orderings. A defining characteristic of the corporate food regime has been the penetration of private capital into the public sphere through mechanisms such as public–private partnerships. The result, as evidenced most clearly by the Bt brinjal case, is that boundaries between public and private interests are blurred and it becomes more difficult for the state to act in the public interest.

In all three cases examined in this book, corporations have used private intellectual property (IP) instruments to bypass existing safeguards in public laws regulating IP rights in agriculture. In Brazil, the private IP regime was implemented in the form of private contracts among the different players in the soybean industry for testing soybeans and charging royalties when farmers sell their harvests to the grain elevator. That system effectively prevented farmers from freely saving seeds for replanting. In India, a private IP regime was implemented more upstream in the soybean production chain, through sublicensing agreements with the seed companies producing Bt cotton. These agreements stipulated that the Bt trait could only be introgressed in hybrid varieties whose seeds cannot be efficiently saved for replanting. In both countries, these private IP systems ensured that Monsanto was able to exact high rates of royalties, regardless of whether the corporation held a valid patent (in the case of Roundup Ready 1 soybean) or had even applied for one (in the case of BG-I Bt cotton). Most important, these private IP systems ensured that farmers could not exercise the right to freely save and replant seeds guaranteed under the domestic plant variety protection (PVP) legislation. In their legal challenges, litigants questioned this subordination of constitutional rights and public law to private contract law. In Brazil, farmers' unions argued that a private contract signed between Monsanto and the producers' federation, without farmers' having been consulted, could not deprive those farmers of their statutory rights. In India, the private arrangements signed with seed producers were even further removed from farmers, yet the cost of royalties was ultimately passed on to farmers in the high price of Bt cotton seeds. Moreover, Brazilian and Indian litigants alike argued that the state has an obligation to guarantee the constitutional rights of its citizens, and therefore cannot recuse itself from intervening in these disputes on the grounds that the matter is the object of a private agreement.

Second, these legal cases—in particular, the Roundup Ready soybean class action lawsuit—sought to revert the three-decade-old trend toward the strengthening of property rights over plant varieties and to establish the right to save seeds as a fundamental right. Increasing restrictions on age-old seed-saving practices have been accompanied by a subtle yet disturbing shift in legal language, with farmers' rights to seeds increasingly couched as "privileges" and "exceptions" being subordinated to the dominant "rights" of plant breeders. There is an irony here for, as Susan Sell (2003, 5) reminds us, "IP rights used to be considered 'grants of privileges' that were explicitly recognized as exceptions to the rules against monopolies." With the shift toward proprietary seeds, the original meaning of "farmers' rights to save seeds" is being distorted, diluted, and ultimately lost. It is not uncommon to hear that the collection of royalties on grains harvested from saved seeds is "nothing but a due charge," since the farmer has in fact not paid royalties on saved seeds. This is, however, a distortion of the original meaning of farmers' rights. When farmers buy seeds from a protected variety, they pay royalties embedded in the price of that seed and, in exchange, they have the unfettered right to save seeds from that protected variety.[2] At the transnational level, the inclusion of rights to seed and to biological resources, as stated in the Declaration on the Rights of Peasants and Other People Working in Rural Areas, adopted by the United Nations in 2018, must be understood in this context (United Nations 2018). By incorporating these rights in the international human rights framework, agrarian movements along with their allies aim to revert the trend toward the marginalization of seed-saving practices and reassert the primacy of both individual and collective rights to seeds over trade and IP rights.

Third, these legal cases raised the thorny issue of the conflict between PVP law and patent law when it comes to biotech seeds. Monsanto has invariably argued that biotech traits fall under the exclusive protection of patent law, an interpretation upheld by the Supreme Courts of both Canada and the United States. In Brazil and India, where the patent and PVP legislation differ significantly from those of the United States and Canada, this conflict had not been addressed until recently. As I have detailed, Brazilian and Indian courts have wavered in their approach to this issue. Like its US and Canadian counterparts, the Brazilian Superior Court of Justice solved the conflict in favor of patent law. In contrast, India's Delhi High Court held that plant cells and seeds are explicitly

excluded from patentability under the Indian Patents Act and that transgenic seeds or plants therefore fall under the purview of the Protection
of Plant Variety and Farmers Rights Act and its research and seed-saving
exemptions.[3]

Fourth, the Delhi High Court came closest to addressing the more fundamental issue at stake in these legal disputes: that of the patentability
of genes and plants. In its ruling, the high court questioned whether biotech traits—described in patent applications as nucleic acid sequences—
could qualify as microorganisms patentable under the WTO Agreement
on Trade-Related Aspects of Intellectual Property Rights (TRIPS). This ties
back to the ambiguous wording of Article 27.3(b) and to the lack of an
agreed-on definition of what constitutes a microorganism. Breaking with
the tendency of patent offices and courts to gloss over these discrepan-
cies, the High Court held that a nucleic acid sequence is *not* a microscopic
organism because it has no existence of its own and has no usefulness
unless it is introgressed into plants, which are not themselves patentable.

While the Delhi High Court had the merit of addressing the more
complex issues surrounding biotech patents, it has only scratched the
surface. Indeed, patent law is increasingly out of step with the state of
the art of scientific knowledge. Social scientists have shown how the
reduction of a gene to a chemical compound that could be isolated has
been instrumental in turning it into something considered patentable
(McAfee 2003; Calvert and Joly 2011). Scientific advances over the past
twenty years, however, have made this convenient simplification increasingly untenable. As epigenetics and postgenomics[4] reveal the complex
expression and regulation of genes, proteins, and their interactions with
cells and organisms, they have rendered obsolete earlier conceptualizations of the gene, to the extent of questioning the concept of "gene"
as a meaningful ontological category (Calvert and Joly 2011). Yet in a
compelling example of discontinuities and ruptures among knowledge
domains (Lock 2005), patent law continues to be based on reductionist
conceptualizations of the gene. It is high time that IP law takes stock of
these scientific developments and fully considers their implications for
the patentability of life forms.

Taken together, these legal cases challenge the legitimacy of the
TRIPS global IP regime in agriculture and contribute—even if only

incipiently—to the emergence of a new legal "common sense" concerning the patentability of seeds and plants (Souza Santos 2002). These cases illuminate the deeply political nature of patents. They also show how legal disputes around IP regimes for biotech crops in the Global South are forging the development of alternative legal interpretations of the balance between, on the one hand, property rights and, on the other, individual and collective rights to seeds. I fully concur with Shobita Parthasarathy (2017) that despite sustained efforts and pressures on the part of transnational corporations and their governments to further harmonize IP laws, true harmonization is likely to be impossible insofar as life forms are concerned.

The agricultural IP landscape is clearly in flux, and I will end by outlining emerging trends that will inflect future conflicts around IP and seeds. One major trend is the unabated pressure on countries to amend their domestic legislation so that plant breeders' rights become akin to patents. By joining the 1991 Act of the UPOV Convention, a country reinforces the exclusive rights of plant breeders—for example, by extending these rights to the product of the harvest. In practice, this would institutionalize the private royalty collection systems implemented in countries like Brazil. Moreover, under UPOV 1991, a plant can be protected simultaneously by a patent and by plant breeders' rights. In practice, this "dual protection" means that the more extensive protection granted by a patent ends up prevailing and thereby nullifying the exemptions under plant breeders' rights. These legislative changes would foreclose the possibility of arbitrating the conflict between patent law and plant variety protection law in favor of the latter and would also prevent the kind of legal challenge mounted by farmers' unions in Brazil.

The looming expiration of the first major agbiotech patents in agriculture—notably of the last foundational patent covering the Roundup Ready herbicide tolerance trait, in 2014[5]—prompted much speculation around what form a "post-patent era" might take. If the fierce fight around patents on new gene-editing technologies is any indication, patents will remain an important tool for industry to appropriate and profit off new technologies (Egelie et al. 2016; Montenegro de Wit 2020). Gene or genome editing (GE) refers to a range of techniques increasingly used since 2015 to alter the genetic material of plants, animals, and other organisms. The most commonly used GE technique, known as CRISPR Cas9, relies on the inherent ability of Cas9 enzymes to cleave foreign DNA (part of a bacteria's immune system). In contrast to earlier recombinant DNA techniques, whereby a trait is randomly inserted in the genome of a living organism, GE allows the insertion, deletion, or substitution of DNA sequences at specific sites in the genome.[6] Genome editing could significantly expand the scope of proprietary rights in plants. Indeed, by overcoming a cell's protective mechanisms, genome editing opens access to the whole genome. Genome editing also expands the types of edits possible (knocking out, activating, silencing, altering, enhancing, deleting, or inserting) and increases the array of organisms in the agroecosystem that can be modified using genome editing techniques (animals, plants, soil microbes, insects).[7] With genome editing, all of these become putatively subject to IP, leading some analysts to argue that there needs to be fundamental redesign of the IP system for plant innovation toward openness rather than exclusivity (Kock 2021).

Moreover, as the private IP systems discussed here make abundantly clear, it is imperative to look beyond public orderings through patents and plant variety protection. Indeed, the industry is actively devising new marketing strategies, including the use of regulatory data and approvals to retain control over its products in what has been labeled "IP-regulatory complexes" (Jefferson and Padmanabhan 2016; Marden, Godfrey, and Manion 2016). Even if a patent expires, the original patent holder can retain a significant level of market control through ownership of the regulatory data required to obtain approval. In fact, the cost of generating new data and documentation means that it is often cheaper for other companies to obtain a license from the original patent owner for an already-approved trait, even if the technology has entered the public domain.

Finally, a further development of relevance is the dematerialization of plant genetic resources—that is, the digitalization of genetic sequences and their storage in electronic databases. This raises daunting new challenges, since the international instruments in the matter of equitable access to genetic resources and benefit sharing have been built around access to and the circulation of material seed samples. Their dematerialization renders these mechanisms obsolete and makes it more difficult to trace these resources back to the communities that cultivate them, therefore opening the door to biopiracy on a new scale.

Given the rapid pace at which biotechnology and IP regimes are coevolving, the challenge for farmers and agrarian activists alike will be to stay ahead of the game. One thing is certain: after over a decade of legal activism around intellectual property and biotech crops, they have become increasingly law-savvy.

APPENDIX A: TIMELINE OF THE COURT CASES

RR Soybean Class Action	
April 2009	Passo Fundo rural union files a class action against Monsanto
April 2012	First-instance decision in favor of rural unions
June 2012	Superior Court of Justice rules in favor of the rural unions in the Special Appeal
Sept. 2014	Second-instance decision in favor of Monsanto
Oct. 2019	Superior Court of Justice rules in favor of Monsanto
Bt Cotton Trait Fee Dispute	
Nov. 2015	Monsanto terminates Nuziveedu's sublicensing contract
Dec. 2015	Ministry of Agriculture issues Cotton Seeds Price (Control) Order
Feb. 2016	Monsanto files lawsuit against Nuziveedu for patent infringement
	Competition Commission issues report in antitrust investigation on Monsanto
May 2016	Ministry of Agriculture issues Draft Licensing Guidelines and Formats for GM Technology Agreements
March 2017	Delhi High Court delivers first ruling in patent infringement lawsuit

(continued)

April 2018	Delhi High Court delivers second ruling, revoking Monsanto's patent
Jan. 2019	Supreme Court of India sends the case back to Delhi High Court for reexamination
April 2021	Bayer-Monsanto and Nuziveedu reach out-of-court settlement
Bt Brinjal Biopiracy Case	
Feb. 2010	National moratorium on the commercialization of Bt brinjal
Nov. 2012	ESG files a public interest lawsuit (PIL) in Karnataka High Court (HC)
	NBA/KBB file a criminal complaint against UAS Dharwad in the Karnataka HC
Oct. 2013	Karnataka HC rules in favor of NBA/KBB in the criminal complaint
Dec. 2013	Karnataka HC transfers the PIL to the National Green Tribunal (NGT)
Feb. 2014	Supreme Court grants a stay on decision of the Karnataka HC in appeal filed by the accused (still in effect in 2021)
March 2014	ESG files a Special Leave petition in the Supreme Court challenging the transfer of the PIL to the NGT (no ruling as of 2021)

APPENDIX B: LOG OF INTERVIEWS

No.	Name or Descriptor, Location	Date
#1	Academic, IP and competition law, Delhi	November 17, 2015
#2	Academic, Munich IP Law Center (Skype)	December 23, 2015
#3	Academic, Center for Policy Research, Delhi	January 15, 2016
#4A	Competition lawyer, Delhi*	January 26, 2016
#4B	Competition lawyer, Delhi*	June 4, 2016
#5A	Academic, economics, Delhi	January 29, 2016
#5B	Academic, economics, Delhi	February 12, 2016
#6	Legal researcher and activist, Delhi	February 12, 2016
#7A	Leo Saldanha and Bhargavi Rao, Coordinators ESG, Bangalore	March 31, 2016
#7B	Leo Saldanha, coordinator, ESG, Bangalore	March 20, 2017
#8A	Seed industry representative (retired), Delhi	April 28, 2016
#8B	Coordinator, Organic Seed Initiative, Biodynamic Association of India, Delhi	April 30, 2016
#9	Suman Sahai, chairperson, Gene Campaign, Delhi	May 3, 2016

(continued)

No.	Name or Descriptor, Location	Date
#10A	Academic, industrial policy and patent law, Delhi	May 4, 2016
#10B	Academic, industrial policy and patent law, Delhi	March 13, 2018
#11	Activist, Peasant Confederation/LVC, Geneva	May 16, 2016
#12	Journalist, Delhi	May 31, 2016
#13	Academic, international environmental law, Delhi*	June 2, 2016
#14A	Food sovereignty activist, Bangalore*	June 5, 2016
#14B	Food sovereignty activist, Delhi	March 16, 2018
#15	Criminal lawyer, Bangalore*	June 7, 2016
#16A	Journalist, rural India, Delhi	June 7, 2016
#16B	Journalist, Rural India, Delhi	March 12, 2018
#17	Journalist, agricultural and environmental issues, Delhi	June 7, 2016
#18	Academic, environmental economics, Dharwad*	June 10, 2016
#19	Administrative officers (joint interview), UAS Dharwad, Dharwad*	June 10, 2016
#20	Leo Saldanha, coordinator, ESG, Bangalore*	June 13, 2016
#21	Lawyer, seed industry, Hyderabad	June 20, 2016
#22	NGO, environment and agriculture, Hyderabad	June 21, 2016
#23	NGO, sustainable agriculture, Hyderabad	June 21, 2016
#24	Plant scientist, ICRISAT, Hyderabad	June 22, 2016
#25A	Executive director, seed industry association, Delhi	June 24, 2016
#25B	Executive director, seed industry association, Delhi	March 12, 2018
#26	Chairperson, Indian farmers' organization, The Hague	October 16, 2016

No.	Name or Descriptor, Location	Date
#27	Agronomist, EMBRAPA, Brasília	November 1, 2016
#28	Legislative consultant, agricultural policy, Chamber of Deputies, Brasília	November 11, 2016
#29A	Néri Perin, lawyer for the farmers' union, Brasília	December 2, 2016
#29B	Néri Perin, lawyer for the farmers' union, Brasília	February 22, 2017
#29C	Néri Perin, lawyer for the farmers' union, Brasília	November 20, 2017
#30	Academic, agricultural research center (Skype)	December 16, 2016
#31	Legal consultant, IP and agriculture (Skype)	January 16, 2017
#32	Former Minister of the Environment, Delhi*	January 25, 2017
#33	Luiz Fernando Benincá, soybean producer, union leader and litigant, Passo Fundo (RS)	January 25, 2017
#34	Soybean producer and union leader, Passo Fundo (RS)	January 26, 2017
#35	Soybean producer and union leader, Passo Fundo (RS)	January 26, 2017
#36	Soybean producer and union leader, Florianópolis (SC)	January 27, 2017
#37	Plant scientist, Camboriú (SC)	January 30, 2017
#38	Lawyer, family farmers' union, Brasília	February 18, 2017
#39	Academic, plant scientist and expert witness, Brasília	February 21, 2017
#40A	IP lawyer, Delhi*	February 21, 2017
#40B	IP lawyer, Delhi*	February 25, 2017
#41	Lawyer, international environmental law (Skype)	March 2, 2017
#42	Academics (joint interview), IP and biotechnology, Delhi	March 14, 2017
#43	Academic, agricultural policy, Delhi	March 15, 2017
#44	Director and horticultural scientist, Karnataka Department of Horticulture, Udupi	March 17, 2017

(continued)

No.	Name or Descriptor, Location	Date
#45	Producer and manager, Mattu Gulla Growers Association, Udupi	March 17, 2017
#46	Panchayat president, Udupi	March 18, 2017
#47	Trustee, rural development NGO, Udupi	March 18, 2017
#48	Mattu Gulla farmer, Mattu, Udupi	March 18, 2017
#49	Temple administrator, Sri Krishna Matha, Udupi	March 18, 2017
#50	Ex-chairman, Karnataka Biodiversity Board, Bangalore	March 21, 2017
#51	Ex-president, biotechnology industry association, Bangalore	March 22, 2017
#52A	Litigant (lead petitioner), Mhow	March 23, 2017
#52B	Litigant (lead petitioner), Delhi	March 18, 2018
#53	Deputy and former president, FETAG, Porto Alegre (RS)	May 16, 2017
#54	Federal judge, Porto Alegre (RS)	May 16, 2017
#55	Family farmer and union leader, Três Passos (RS)	May 18, 2017
#56	Soy grower, Três Passos (RS)	May 18, 2017
#57	Soy grower, Três Passos (RS)	May 18, 2017
#58	Family farmer and union leader, Ijuí (RS)	May 19, 2017
#59	Soy grower, Ijuí (RS)	May 19, 2017
#60	Soy grower, Ijuí (RS)	May 19, 2017
#61	Family farmer and union leader, Santo Anjo (RS)	May 20, 2017
#62	Soy grower, Coronel Barros (RS)	May 20, 2017
#63	Family farmer and union leader, Coronel Barros (RS)	May 20, 2017
#64	Soy grower, Coronel Barros (RS)	May 20, 2017
#65	Federal deputy, Workers' Party, Brasília	August 30, 2017
#66	Civil servant, Ministry of Agriculture, Brasília	January 8, 2018

No.	Name or Descriptor, Location	Date
#67	IP lawyer involved in the Bt cotton lawsuit, Delhi	March 14, 2018
#68	Leader, RSS organization, Delhi	March 14, 2018
#69	Ex-official, NBA (phone)	March 15, 2018
#70	Legal counsel and technical/regional officer (joint interview), UPOV, Geneva	April 12, 2019
#71	Activist (IP and agriculture), NGO, Geneva	April 12, 2019
#72	Patent Law Division, WIPO, Geneva	April 17, 2019
#73	Agronomist, EMBRAPA, Brasília	May 6, 2019
#74	USDA Foreign Agricultural Service, Brasilia	May 6, 2019
#75	International relations / R&D researchers (joint interview), EMBRAPA, Brasília	May 7, 2019
#76	Scientist, plant biotechnology, EMBRAPA, Brasilia	May 7, 2019
#77	IP adviser, EMBRAPA, Brasilia	May 7, 2019
#78	Lawyer with STJ expertise (email)	October 1, 2019

* Note: Interviews conducted by a research assistant are marked with an asterisk. Letters indicate follow-up interviews.

APPENDIX C: BT BRINJAL AND THE INTERNATIONAL REGIME GOVERNING PLANT GENETIC RESOURCES

The Nagoya Protocol to the Convention on Biological Diversity (CBD) and the FAO Plant Treaty have seldom been mentioned in connection with the Bt brinjal case. Trying to come to grips with their relevance and implications in this specific instance sheds light on the complexity and ambiguities of the international legal regime governing access to genetic resources. Do the local eggplant varieties used in the development of Bt brinjal come under the international regime? Who makes decisions about these varieties? Should local communities have provided their informed consent for the use of these local varieties in the development of Bt brinjal? And, finally, were they entitled to benefit sharing?

The CBD and the Plant Treaty adopt distinct approaches to access and benefit sharing. Under the CBD, bilateral contracts are negotiated between the provider and the user of a genetic resource, in accordance with the national legislation implementing the CBD. In contrast, the Plant Treaty creates a multilateral system for facilitated access to plant genetic resources for food and agriculture as well as for equitable benefit sharing at standard, uniform conditions that are agreed on internationally. Material available under the multilateral system is accessed through the Standard Material Transfer Agreement (SMTA) that specifies the rights and obligations of each party (that is, the individual providers and recipients entering the contract). The multilateral system covers genetic material from 35 food crops (eggplant is one of them) and 29 forage crops. Plant genetic resources that are not within the scope of the multilateral system of the Plant Treaty may be accessible under the national legislation, if any, implementing the CBD/Nagoya Protocol.

In the case of Bt brinjal, the key issue is to determine the legal status of the local eggplant varieties used. According to two specialists of the Plant Treaty, as plant genetic resources under the control and management of the state and in the public domain (that is, free from IPR claims), the eggplant varieties, at least in theory, should have been in the multilateral system (Interviews #30 and #41). In practice, however, this is one of the Plant Treaty's gray areas. Whether germplasm kept by public agricultural universities comes under the management and control of the State or is an institution's private property varies from country to country. In India, the legal status of germplasm held by public universities is not defined (Interview #41). The university could have chosen to submit an SMTA, but chose not to on the grounds that this material belonged to the university. It could be argued that the university had invested in maintaining and possibly developing this material, and could therefore make some claims over it. However, as the university asserted that this material was its property, the local community that had provided the material in the first place found itself excluded from any rights over it and from any claims to benefit sharing.

A political dimension complicated the university's decision. At the time, India had not yet legislated the Plant Treaty and designated which germplasm collections would be included in the multilateral system (this was done in 2014). Like a number of other countries, India was reluctant to do so because it felt that the country had already contributed its share of plant genetic resources to the multilateral system through the international network of agricultural research centers. In the absence of clear rules and political backing, Indian institutions felt wary of taking responsibility for including national resources in the multilateral system (Interview #30).

In a hypothetical scenario in which the public universities would have followed the rules of the multilateral system, the transfer of these varieties would have been accompanied by an SMTA. The consortium could have obtained a patent on Bt brinjal, but a percentage of commercial sales

(0.77 percent) would have reverted to the Plant Treaty's Benefit-Sharing Fund. If the universities had chosen to go the CBD/Nagoya route, they would have provided for prior informed consent through the National Biodiversity Authority and entered into a bilateral contract with the public–private consortium. The consortium would have, in turn, required the informed consent of the local communities that had provided these varieties to the universities, and dealt with provisions on benefit sharing with these communities.

The universities chose a third route by signing ad hoc agreements with the consortium, in spite of existing, applicable national legislation. Such agreements do not follow either the bilateral logic of the Nagoya Protocol or the multilateral logic of the Plant Treaty. In fact, they fall outside the purview of the international legal regime and entail no obligations with regard to informed consent and benefit sharing.

APPENDIX D: UDUPI MATTU GULLA: A CASE STUDY

Of the six varieties of eggplant (*brinjal*) used to genetically engineer the Bt trait, at least one is well documented. Udupi Mattu Gulla is a round, light-green eggplant variety renowned for its unique taste and exclusive locale of production. *Gulla* means eggplant in the local language, Tulu. Approximately 200 farmers cultivate this variety on 70 hectares adjacent to the small village of Mattu, located in Udupi district, in coastal Karnataka, approximately 400 kilometers from the State capital, Bangalore. This local variety has deep historical, cultural, and religious significance, being closely associated with a 400-year-old religious festival, the Paryaya. The festival is celebrated on January 18 in alternate years at the Sri Krishna Hindu temple in Udupi, and it marks the transfer of the temple administration.

A temple administrator tells the legend surrounding the origins of the Mattu Gulla variety:

Earlier, there was a priest, *Swami Raja*, he worshipped the God *Hayagriva* and used to give him *Prasada* [a food offering to the gods] made of Bengal gram. But some of the devotees felt that this particular *Swami* (priest) was consuming everything and not giving it to the God *Hayagriva*. Once they suspected that, they put poison into the *Prasada*. But instead of the priest dying, the God became blueish. The devotees then realized that Swami-ji did not consume the *Prasada*, God did. The devotees felt that what they had done was wrong. They asked the Swami-ji to give them a remedy. As a remedy, the Swami-ji gave a

A.1 Mattu Gulla eggplant, Udupi, Karnataka, India

particular kind of seeds to the people of Mattu and asked them to grow that vegetable, and to offer *Hayagriva* food made of that vegetable. When the God consumed the Prasada made of Mattu Gulla brinjal, the blue poison vanished from the God's body except for a small spot near the throat. (Interview #49; as told by a temple administrator in Kannada and translated into English by an interpreter)

The fruits of the Mattu Gulla are round in shape, light green in color, with white stripes. The small spines on its stalks are said to confer some protection against the fruit and shoot borer insect, because it prevents it from laying its eggs. Mattu Gulla is appreciated for its low astringency and thin skin that dissolves in cooking.

Local people say that while it can be grown in other places, Mattu Gulla cultivated in Mattu has a unique taste. The local people attribute these characteristics to the unique soil and climate conditions. Mattu Gulla is cultivated on 150 to 200 hectares of land between the Indian Ocean and two rivers (the Mattu and the Papanashini). Traditionally, farmers (who are also usually fishermen) used fish manure as compost, but the price has gone up, so they now use organic compost instead. When the land is immersed in the rainy season, some farmers also grow paddy rice. Eggplant is planted after the monsoon, in September (this region of India is the first to receive the southwest monsoon), and harvested every two weeks from October until March. Part of the first crop is brought to the temple. Another part is used as seeds (no commercial seeds are available, so farmers produce their own seeds). Finally, the remainder is sold on the market.

Farmers found out that the Mattu Gulla variety had been used to develop Bt brinjal when the Indian government approached agricultural research institutes in Karnataka to identify farmers who would be willing to test Bt brinjal in the fields (Interview #45). This prompted local farmers to form an association, the Mattu Gulla Growers' Association, which has some 150 members. With the support of the State Department of Horticulture, it applied to have Mattu Gulla recognized as a Geographical Indication (GI). A GI is attributed to a product whose qualities and characteristics are intrinsically linked to its place of production. The Mattu Gulla Growers' Association obtained the status of Geographical Indication (GI Certificate no. 199) in 2011. Significantly, the movement to have the Mattu Gulla eggplant variety declared a GI stemmed from the concern that it had been genetically engineered without the community's knowledge or consent (Bhat and Madhyastha 2007). This initiative has had a positive impact on farmers: by cutting out the middlemen, they received a better price for their vegetables. However, it is important to stress that the GI status protects the commercial use of the name, not the physical resource itself, and therefore does not tackle issues surrounding access and benefit sharing (Interview #30). Why the developers of Bt brinjal chose such a unique variety is open to question. According to one horticultural scientist, it could have been chosen for its spines, which confers some resistance to the fruit and shoot borer, as well as for its unique taste (Interview #44).

NOTES

INTRODUCTION

1. English translations of Brazilian proper nouns are followed by their Portuguese acronyms. See Acronyms and Abbreviations for the full Portuguese names.

2. Each social movement has its own banner, anthem, and insignia: small farmers, for example, use a straw hat, landless peasants a red cap, and peasant women a mauve scarf.

3. In 2000, AstraZeneca and Novartis merged their agrichemical businesses to form Syngenta AG, which was acquired by China National Chemical Corporation (ChemChina) in 2015.

4. I use the term *seed* broadly to include every plant structure (for example, seeds, seedlings, cuttings) used in the propagation of a plant.

5. Critically examining the proprietary dimension of biotech crops is all the more important, given the industry's "pro-poor technology" narrative. See Stone (2002) and Glover (2010a, 2010b).

6. "Proprietary seeds" are branded seed varieties subject to intellectual property protection.

7. These four corporations continue to consolidate their hold on the market even further by entering into agreements over the cross-licensing of IP, research, and development. See IPES-Food (2017).

8. A plant variety meets the DUS criteria for plant variety protection if it differs from existing varieties ("Distinct"), if its characteristics are expressed uniformly ("Uniform"), and if those characteristics do not change over subsequent generations ("Stable"). "New" in this context means that it is new on the market.

9. For a more detailed explanation of the differences between traditional and molecular plant breeding, see Krimsky (2019), chapters 1–3.

10. In an 1889 decision rejecting a patent application on a pine fiber, the US Patent Office wrote: "If [such a patent] were allowed, patents might be obtained upon the trees of the forest and the plants of the earth, which of course would be unreasonable and impossible" (cited in Beauchamp 2011, 13).

11. For a more detailed discussion of these cases, see Aoki (2008), Pollack (2004), and Peavey (2014).

12. The Supreme Court of Canada judgment in *Monsanto v. Schmeiser*—a 2004 decision narrowly grounded in patent law—failed to consider other important questions related to biosafety, environmental liability, and farmers' rights. Indeed, the judgment was widely criticized as relieving companies of any responsibility or liability for genetic contamination (Cullet 2005a). The Supreme Court determined that it was up to Parliament to consider these issues and to amend the Patent Act accordingly.

13. Out of 147 patent violation suits it has filed against US farmers since 1997, Monsanto won all 9 cases that went to trial (Schapiro 2018). In the remaining cases, Monsanto reached confidential, out-of-court settlements.

14. In addition to intellectual property, other mechanisms contribute, directly or indirectly, to the enclosure of seeds, notably seed laws (which regulate seed production, marketing, and trade) and phytosanitary regulations (which set plant health standards and food safety rules). See Wattnem (2016).

15. The corporate food regime was preceded by a colonial-diasporic food regime (1870–1930s) and a mercantile-industrial food regime (1950s–1970s).

16. *Evergreening practices* are various legal, business, and technological strategies used by industry to extend the duration of patents that are set to expire and thereby to prevent inventions from entering into the public domain. *Patent thicket* refers to the dense web of overlapping IP rights that can cover an invention, thus requiring competitors to enter into multiple licensing deals.

17. Rosemary Coombe stands out for addressing intellectual property and human rights in a 1998 article. See Coombe (1998).

18. See Canfield (2020) on ethnographic approaches to property; and Coombe and Chapman (2020) on intellectual property more specifically.

19. Thanks to an anonymous reviewer for this wording.

20. On Argentina, see Lapegna and Perelmuter (2020); on Colombia, see Escobar and Fitting (2016); on Pakistan, see Rana (2021).

21. The Global South is understood here as a geopolitical concept and as a political subjectivity, which encompasses people as well as spaces that are in a subaltern position in relation to global capitalism (Mahler 2018).

22. In 2019, the top producers were the United States (72 million hectares, or Mha), Brazil (53 Mha), Argentina (24 Mha), Canada (13 Mha), and India (12 Mha) (ISAAA 2019).

23. One hectare is equivalent to 2.47 acre, or 10,000 square meters. The International Service for the Acquisition of Agri-Biotech Applications (ISAAA) is an industry

group actively promoting GM crops worldwide. Its annual statistics on GM crop production are widely cited, as the organization is the only available source of statistics on GM crops globally. Its data, however, must be used with caution. ISAAA provides statistics for countries for which there are no official statistics; it does not disclose the source of its information; and its statistics have sometimes been found to be inflated. See FOEI (2006).

24. The family farming sector produces an estimated 70 percent of the food consumed domestically (MDA 2008, 5).

25. According to the country's 2015–2016 census of agriculture, there are close to 126 million small and marginal farmers in India, who each own less than two hectares of land. They account for 86 percent of all farmers but own just 47 percent of the total crop area (GoI 2020).

26. Public agricultural research in Brazil is conducted under the authority of the Brazilian Agricultural Research Corporation (EMBRAPA), and in India it falls under the Indian Council of Agricultural Research (ICAR).

27. In India, the National Gene Bank holds over 395,000 samples, representing 1,584 species (NBPGR n.d.). In Brazil, EMBRAPA Genetic Resources and Biotechnology—part of the Brazilian Agricultural Research Corporation—holds 140,000 seed samples, representing 960 species (EMBRAPA n.d.).

28. Both countries are parties to the CBD. India ratified the Plant Treaty in 2003, and Brazil did so in 2008. The Plant Treaty came into force internationally in 2004. India ratified the Nagoya Protocol in 2012, and Brazil in 2021.

29. According to GRAIN and LVC (2015, 28), "As early as 2003, Monsanto had a department of 75 employees with a budget of 10 million USD dedicated to the sole purpose of pursuing farmers for patent infringement."

30. In 2015, the Big 6 were BASF, Bayer, Dow Chemical, DuPont, Monsanto, and Syngenta. By 2018, these companies had merged to become a trinity of DowDuPont, Bayer-Monsanto, and ChemChina-Syngenta.

CHAPTER 1

1. The role of transnational corporations in formulating the WTO Agreement on Trade Related Aspects of Intellectual Property Rights (TRIPS) has been amply documented (Matthews 2002; Sell 2003, 2009). As a pharmaceutical industry representative candidly said at the time, "industry . . . crafted a solution, reduced it to a concrete proposal and sold it to our own and other governments" (Oh 2000).

2. The TRIPS Agreement was unprecedented in that it established a link between two issues—trade liberalization and intellectual property—previously seen as having no logical connection (Purdue 2000).

3. Canada, for example, enacted its first plant breeders' rights legislation in 1990, and Norway in 1993.

4. This section is based on a comparative reading of accounts by two of the main negotiators for Brazil and India, Tarragô (2015) and Ganesan (2015), respectively.

While the pharmaceutical industry loomed large in the negotiation of the TRIPS Agreement, I focus in this section on the discussions related to plant biotechnology.

5. The Special 301 Report of the Omnibus Trade and Competitiveness Act 1988 is published annually by the Office of the US Trade Representative as a unilateral measure to put pressure on countries to increase IP protection beyond what is required by the TRIPS Agreement.

6. A cosmid is a type of hybrid plasmid (DNA molecule) that allows cloning of large DNA fragments.

7. This strategy was well captured in the title of a 1996 article by GRAIN: "UPOV: getting a free TRIPs ride?" (GRAIN 1996). On TRIPS-compatible alternatives to UPOV, see Helfer (2004) and Correa et al. (2015).

8. The Brazilian Industrial Property Code of 1945 provided for the possibility of protecting IP in plant varieties, but it depended on the enactment of special regulations, which never occurred.

9. At the time, Brazil was in the throes of a dictatorship (1964–1985). Without downplaying the role played by mobilization in defeating the bill, the military government's national-developmental ideology and the fact that it saw food security as a strategic issue also played a role in the bill's demise (Pelaez and Schmidt 2000).

10. Founded in the aftermath of the military regime (1987), IDEC is Brazil's oldest and largest consumer protection organization. The Brazilian chapter of Greenpeace International was created in 1993.

11. The PVP Act defines a small rural producer as someone who exploits a parcel of land using mainly family labor as opposed to hired labor, and who resides on their property or at least nearby. To avoid the inclusion of unproductive large estates (*latifúndios*), properties in this category may not exceed size limits set forth in the Act (RFB 1997, Art. 10.3).

12. Since 1999, countries that choose to join UPOV no longer have the option to do so under the 1978 Act—they must obligatorily join under the 1991 Act.

13. President Collor de Mello was impeached later that year, following corruption charges.

14. Fernando Henrique Cardoso was president from 1995 to 2002.

15. Farmers are defined in the PPVFR Act as persons who cultivate crops themselves or through direct supervision or who conserve and add value to wild species or traditional varieties through selection and identification of their useful properties (GoI 2001, Art. 2k).

16. Ethiopia, the Philippines, and to lesser extent Malaysia have similar provisions.

17. In the first 10 years, the PPVFR Authority issued 3,538 certificates of registration for plant varieties. Of those, 44 percent were for farmers' varieties. Data compiled from PPVFR Authority (2019).

18. Section 3 of India's Patents Act excluded from patentability "any process for the . . . treatment of animals or plants to render them free of disease or to increase their economic value or that of their products" (GoI 1970).

19. The case of India is instructive: if it were not for pressure from civil society actors, India would likely have joined UPOV and agreed to plant variety protection norms based on UPOV 1978. This was the path taken by most countries that lacked either a strong civil society or the resources to develop their own sui generis legislation, and thus felt more vulnerable to external political and economic pressures. UPOV had only some 20 members in the early 1990s; by 2017, it had more than tripled its membership to 75 countries (GAIA/GRAIN 1998; UPOV n.d.).

20. This is referred to in the literature on globalization as "cunning states" (Randeria 2003b) or "bounded autonomy" (Newell 2006).

21. Under US law, only laws of nature, natural phenomena, and abstract ideas are considered ineligible for patent protection.

CHAPTER 2

1. The term *event* designates the insertion of a particular transgene into a specific location on the chromosome and is used to identify genetically engineered crop varieties.

2. Unless otherwise indicated, all translations from the Portuguese are my own.

3. As one lawyer observed in a letter to the Antitrust Division of the US Department of Justice, "Probably the most well-known examples of patents covering the use of off-patent agrochemicals on GMOs are the patents governing the use of the herbicide Glyphosate" (Callahan 2009).

4. According to the precautionary principle, if an activity or technology involves potential serious and irreversible threats to human health and the environment, the absence of scientific certainty should not be used to oppose the adoption of precautionary measures.

5. For a more detailed discussion of the controversy around transgenic crops in Brazil, see Pelaez and Schmidt (2004), Scoones (2008), and Motta (2016).

6. This model of private royalty collection system was introduced around the same period in Paraguay (Filomeno 2014).

7. This is one of the key differences between the 1978 Act and the 1991 Act of the UPOV Convention. Under UPOV 91, plant breeders' rights extend to harvested material. Brazil, however, is a party to the 1978 Act.

8. Benincá appealed the decision to Brazil's Superior Court of Justice, but his appeal was dismissed on procedural grounds.

9. Monsanto does not make public how much it collects in royalties, but the sum can be inferred from available data (including sales of certified seeds, total cultivated soybean area, production per hectare, and soybean market prices).

10. Other unions and organizations expressed their interest in backing the lawsuit but were asked not to do so in order to avoid further delays in the proceedings (Interview #29A).

11. The Public Prosecutor's Office (Ministério Público) is a body of public prosecutors at the federal and state levels who are independent of Brazil's three branches of government. They are authorized by the Constitution to bring actions against private individuals, commercial enterprises, the federal state and municipal governments, in defense of minorities, the environment, consumers, and civil society.

12. According to the judge, three years was a reasonable delay, given the complexity of the case, the fact that there was no case law, and the need to analyze patents (Interview #54).

13. In the Brazilian justice system, the Superior Court of Justice (STJ) is the country's highest court for nonconstitutional matters. Litigants can only appeal a decision of the STJ before the Federal Supreme Court if they can demonstrate that the matter in dispute has constitutional relevance.

14. In a number of individual and class actions in Brazil, lower court judges have granted the judicial deposit of royalties. However, higher courts generally overturn these decisions before they become effective. This means that farmers have to continue paying royalties to Monsanto until the final ruling.

15. For example, in 2013, the Soybean Producers Association of the State of Mato Grosso (APROSOJA-MT) filed a case in a Mato Grosso court demanding that Monsanto produce its Brazilian patent on Intacta (Tubino 2013).

16. The corresponding US patent to Brazilian pipeline patent 1100008–2 is US Patent 5627061.

17. Here is a list of the fourteen pipeline patents (in bold are those patents for which Monsanto sought an extension): PI 8706530–4, PI 1100009–0, PI 9007159–0, PI 1100007–4, PI 9007550–1, PI 9508620–0, **PI 1100008–2, PI 1101069–0, PI 1101070–3, PI 1101047–9, PI 1101048–7,** PI 1101049–5, **PI 1101045–2,** and **PI 1101067–3** (Barbosa 2014, 335).

18. In one case (PI 1100007–4, "Glyphosate-resistant plant"), the Brazilian Patent Office conceded the extension but then rescinded its decision (Baumer 2005).

19. This can be done successively. For example, one of Monsanto's patent applications for the glyphosate-tolerance trait in the United States reads: "Continuation of application no. 08/833,485, filed on Apr. 7, 1997, now Pat. no. 5,804,425, which is a continuation of application no. 08/306,063, filed on Sep. 13, 1994, now Pat. no. 5,633,435, which is a continuation-in-part of no. 07/749,611, filed on Aug. 28, 1991, now abandoned, which is a continuation-in-part of application no. 07/576,537, filed on Aug. 31, 1990, now abandoned" (Official Gazette 2001, 2820).

20. This argument is supported by some legal researchers. See Ávila (2015, 121–25).

CHAPTER 3

1. These aspects of the Bt cotton controversy in India are beyond the scope of this book. See Scoones (2006), Herring (2007), Glover (2009), Stone (2012), Kranthi (2016), and Flachs (2019).

2. In India, Monsanto coined the expression "trait fees" to refer to the estimated value conferred by the Bt gene. According to some critics, trait fees are merely "royalties under a new name" (Shiva 2016a).

3. In 1998, Monsanto acquired a 26 percent stake in Mahyco.

4. Monsanto Inc. USA has a wholly owned subsidiary in India called Monsanto Holdings Private Limited. In this chapter, I refer to both as Monsanto.

5. In addition, the sublicense agreement stipulates that the sublicensee must withdraw BG-I Bt cotton three years after the commercial approval of the BG-II varieties, or five years after the first planting of the BG-II varieties (Ramanjaneyulu 2016).

6. Amounts in Indian rupees (INR) are converted to USD using the official exchange rate (LCU per USD, period average) for the year in question.

7. For a short period between 2006 and 2009, Monsanto faced limited competition from three other Indian biotech companies—Metahelix, JK Agrigenetics, and Nath Seeds—which marketed their own first-generation (meaning single-gene) Bt varieties. This competition was short-lived, however, as these companies became Monsanto sublicensees for the two-gene BG-II technology, which, unlike BG-I, was patented in India.

8. The All India Kisan Sabha (or All India Peasant Union) was formed in 1936 as the peasant front of the Communist Party of India (CPI). The Andhra Pradesh Rythu Sangam is also affiliated with the Communist Party of India. The farmers' organizations had the support of civil society organizations, including Gene Campaign and the Indian Social Action Forum.

9. In addition to Monsanto and MMB, the companies cited in the complaint are Mahyco, Nuziveedu, Proagro, and Rasi.

10. Contrary to India, the Chinese Academy of Agricultural Sciences had succeeded in bringing out its own Bt cotton varieties, thus offering stiff competition to Monsanto.

11. In 2007, farmers' organizations successfully pressed the state government to further reduce the price of cotton seeds to 650 INR (16 USD) for BG-I and to 750 INR for BG-II (18 USD) (UNI 2007).

12. The seven states are Maharashtra, Gujarat, Karnataka, Tamil Nadu, Punjab, Madhya Pradesh, and West Bengal.

13. MMB challenged all three states and won against Madhya Pradesh because the state government had failed to enact a special law (Jishnu 2010a).

14. To make up for these losses, the industry changed the recommendation from one packet to two packets of seeds per acre, thus doubling its business (Ramanjaneyulu 2016).

15. Around that period, reports emerged that new forms of child labor were developing in cotton seed-producing regions. Cotton seeds are typically produced on small farms and sold to domestic and multinational companies. These cotton seed producers rely on middlemen to contract adult and child laborers to do the tedious cross-pollination work involved in the production of hybrid cotton seeds (Venkateswarlu 2010). According to one activist, a driving force behind this development was seed companies' quest to maintain their profit margins amid high and unregulated royalties and regulated maximum retail price by squeezing surplus from the labor employed in the production process (Interview #14A).

16. In India, agriculture falls under the jurisdiction of the states, but price controls are the prerogative of the Central Government, thus placing the regulation of cotton seed prices in a gray area between the two levels of government.

17. The RSS's cultural wing, the World Hindu Council (Vishva Hindu Parishad), represents an exception: it is pro-GMOs and denounces critics as "anti-science" (Chowgule 2015). On the divisions among RSS organizations on the issue of GM crops, see Jishnu (2015).

18. The sublicensing agreement stated that "The sub-licensor is empowered to terminate the sub-license agreement with immediate effect if at any time, any laws in the territory restrict the sub-license fees (trait value) specified in Article 3 payable by the sub-licensee to the sub-licensor" (CCI 2016, 16–7).

19. BG-I was initially effective against bollworms, the main pest. However, with bollworms neutralized, other secondary pests such as sap-sucking insects took over and needed to be controlled with pesticides. Moreover, it was only a question of time before the bollworm itself developed resistance. The first evidence that the pink bollworm had developed resistance to BG-I came in November 2009 (Monsanto 2010). In 2006, Monsanto introduced its second-generation cotton technology, called Bollgard-II. Within five years of its introduction, the pink bollworm had also developed resistance to BG-II (Ramanjaneyulu 2016).

20. Under Section 84 of the Patents Act, a compulsory license can be granted if a patented invention is exorbitantly priced, is not available in reasonable quantities, or is not being worked in the territory of India.

21. The ruling was not without its problems. For example, the judges accepted one argument put forward by the National Seed Association of India and instructed Monsanto to apply for protection and benefit sharing under the Protection of Plant Varieties and Farmers' Rights Act (PPVFR Act). The Court suggested that once Monsanto had registered its varieties under the Act, it could seek royalties from other seed companies that use the Bt trait. This line of argument is questionable for two reasons. First, the PPVFR Act allows the protection of a plant variety as a whole, not of specific traits. Second, the benefit-sharing provisions of the Act were intended to compensate farmers and communities whose resources have been used to develop commercial varieties (Reddy 2018b, see also Peschard 2017). To use these provisions to compensate technology providers is altogether a different matter, as well as a distortion of the Act's intent.

22. Indian Patent no. 168950, "A method of producing transformed cotton cells by tissue culture," was granted to Agracetus by the Indian Patent Office in May 1991. The United States revoked Agracetus's patent shortly after India did, on the grounds that the invention did not qualify as novel (Riordan 1994).

23. In the second instance, the Indian government revoked, in 2012, a patent granted by the Indian Patent Office to the company Avasthagen for a medicine to control diabetes based on traditional plants, on the grounds that these properties were already fully documented in the Traditional Knowledge Digital Library.

24. A similar situation has been reported in Pakistan, where it was widely believed in government and business circles that BG-I was patented in that country, and that Monsanto had an international patent on these technologies whereby its patents automatically extended globally (Rana 2021). However, there is no such thing as an "international patent"; a patent is either national or, at most, regional.

25. For example, in an article published in 2006, the Andhra Pradesh Agriculture Commissioner is quoted as saying that Monsanto has no patent on Bt cotton in India (Ramakrishna 2006).

26. Biosafety refers to the set of regulations aimed at preventing the potential risks posed by genetically engineering organisms.

27. This statement is similar to the ones made by Monsanto in Brazil. With regard to Roundup Ready soybean, for instance, Monsanto stated: "RR1 Technology is protected by various types of intellectual property rights, including patent and patent applications, trade and commercial secrets, and regulatory information and approvals, as well as continuing improvements, among others" (FAMATO 2013).

28. That is, 20 years after its international filing date under the Patent Cooperation Treaty.

29. The prevailing consensus is that GM crops simplify farm management but do not increase productivity. See, for example, IAASTD (2009).

CHAPTER 4

1. As reported by Leo Saldanha (Interview #7A).

2. The Bt gene construct also comprises the cauliflower mosaic virus (CaMV) 35S promoter, used to activate the transgene in the host genome, and two antibiotic resistance marker genes, whose function is to identify cells that have been successfully transformed (ISAAA n.d.).

3. This public interest lawsuit (no. 260/2005)—not to be confused with the biopiracy public interest lawsuit discussed in this chapter—was brought to the Supreme Court in 2005 by Aruna Rodrigues to challenge the release of genetically modified organisms (GMOs) by the government of India in the absence of a proper biosafety protocol. As of 2021, the case was still active.

4. Cornell held the patent on this method, which it licensed to DuPont. Monsanto held the patent on the other most common method to induce genetic

transformation in plants: the use of *Agrobacterium tumefaciens*, a bacterium that has the natural ability to infect plant cells. It is this second method that was used in the development of Bt brinjal. Both methods are now in the public domain, although some improvements remain under patent.

5. ABSP-II also sponsored the introduction of Bt brinjal in both Bangladesh and the Philippines. Bt brinjal has been grown commercially in Bangladesh since 2013.

6. A good deal of the controversy over genetically engineered crops in India stems from the conflict of interest due to one committee being responsible for both promoting *and* regulating biotech crops.

7. Ex situ (off-site) refers to the conservation of plant genetic resources in gene banks, as opposed to their conservation in the fields.

8. The Universities of Agricultural Sciences (UAS) are Indian public universities dedicated to agriculture. Dharwad is a city in the northwest of Karnataka. UAS Dharwad is the second oldest UAS in Karnataka, after UAS Bangalore.

9. KIA was announced on the same day as the Indo–US civil nuclear deal. According to Sridhar (2014), it was part of the concessions made by India in the negotiations to arrive at a deal—concessions that would benefit large US corporations.

10. For an early critique of ISAAA's role in Asia, see Kuyek et al. (2000).

11. ESG is a nongovernmental organization formed in 1986 and based in Bangalore, Karnataka. Its small team of ten engages in research, education, and advocacy on issues of environmental and social justice. Over the past two decades, ESG has campaigned on a range of issues, from waste management and tree-felling to the preservation of wetland ecosystems (ESG n.d.).

12. The FAO Plant Treaty was signed in 2001 and came into force in 2004. Its objectives are the conservation and sustainable use of genetic resources for food and agriculture, together with the fair and equitable sharing of the benefits arising out of their use.

13. Jayanthi Natarajan replaced Jairam Ramesh as Environment Minister in July 2011.

14. According to ESG, to put such a decision to a vote was unprecedented and revealed disagreements among the board members (Interview #7A).

15. The CAG's conclusions were supported by a 2012 report on the MoEF by the Public Accounts Committee of the lower house of parliament (Lok Sabha 2012a).

16. On March 10, 2005, Mahyco and Sathguru signed an agreement for research and development for delivery of Bt eggplant for resource constrained farmers. On March 20, 2005, Mahyco entered into a material transfer agreement with TNAU. Finally, on April 2, 2005, Mahyco signed a tripartite sublicense agreement with Sathguru and UAS Dharwad.

17. The six local varieties accessed in Karnataka through UAS Dharwad are listed in the sublicense agreement as Malpur local, Majari Gota, Kudachi local, Udupi local, 112 GO, and Pabkavi local (Mahyco, Sathguru, and UAS Dharwad 2005). The four local varieties accessed in Tamil Nadu through TNAU are MDU I, PLR-1, KKM-1, and CO2 (Mahyco and TNAU 2005).

18. In addition to claims 9–11, Mahyco canceled claim 1, related to "A method of detecting the presence of brinjal plant EE-1 elite event nucleic acid sequence in a sample."

19. In addition to applying for a patent on the transgenic event, Mahyco also applied in December 2010 to obtain plant breeders' rights to the Bt brinjal plant variety under the Protection of Plant Varieties and Farmers' Rights (PPVFR) Act. Plant breeders' rights give the breeder of a new variety exclusive rights over the propagating material of that variety for 15 years. In 2016, according to the website of the PPVFR Authority, Mahyco's application for Bt brinjal was pending due to legal issues (PPVFR Authority 2016).

20. See Appendix C for a discussion of how the issue of access to local eggplant varieties for the development of Bt brinjal relates to the international regime governing plant genetic resources.

CHAPTER 5

1. In 2019, PepsiCo filed an IP infringement lawsuit against Indian potato growers in what is, to my knowledge, the first-ever such lawsuit in India (this lawsuit concerned a plant breeders' certificate, not a patent). The lawsuit was rapidly withdrawn following a spirited defense by the farmers of their rights, a public outcry, and presumably political pressure from the ruling Bharatiya Janata Party (Down to Earth 2019).

2. As for Bt brinjal, Monsanto and its partners in the ABSP-II consortium designed a two-pronged approach, in which the private sector would commercialize hybrid Bt brinjal varieties, while public agricultural universities would pursue the development of open-pollinated varieties. Because of the 2009 national moratorium on GM crops, Bt brinjal had not been commercialized as of 2021. But it is reasonable to assume that, just as with Bt cotton, restricting the Bt trait to hybrids would ensure Monsanto's control over the commercial market for Bt brinjal.

3. Mahyco Monsanto Biotech (MMB) is Monsanto's joint venture in India. See chapter 3.

4. Comment reported by a member of Congress during a session of the Special Commission examining a bill to amend the Plant Variety Protection Act; Brasília, December 5, 2018.

5. This quote echoes the words of the UK's Lord Chancellor, talking about the East India Company in the late nineteenth century: "Corporations have neither souls to be damned nor bodies to be punished. They therefore do as they like." Quoted in Dalrymple (2019).

6. Farmers, however, were represented by a lawyer from the seed company Nuziveedu, which indicates that corporate interests were involved early on (Interview #14B).

7. In negotiating technical cooperation agreements between EMBRAPA (the Brazilian Agricultural Research Corporation) and Monsanto, the former would be represented by one intellectual property adviser, while the latter would send an entire team of lawyers (Interview #77).

8. "Clash of civilization" is an expression reminiscent of Samuel P. Huntington's influential but much-critiqued book whose title begins with these words (1996).

9. Bornstein and Sharma (2016) make a similar observation concerning the India Against Corruption movement, which has allied with Hindu spiritual leaders and used images of Bharat Mata (Mother India) associated with the Hindu right wing.

10. Patent examiners opined that a gene found in nature is *not* patentable, while one that is not found in nature and whose function or utility is specified *is* patentable.

11. *Novartis AG v. Union of India* was a landmark court case on whether the Swiss pharmaceutical company Novartis could patent the drug Gleevec in India. The Supreme Court upheld the Indian Patent Office's rejection of the patent application on the grounds that it did not fulfill the criteria for novelty under Section 3(d) of the country's Patents Act.

12. An ADI can be filed by any of these parties: Brazil's president, the bureau of the Senate or of the Chamber of Deputies, the Federal District governor or the Legislative Assembly, the Office of the Attorney General, the Bar Association of Brazil, a political party represented in Congress, a union confederation or a nationwide class entity.

13. The Attorney General of the Union (Advocacia Geral da União) represents the Federal Union before the national courts and provides legal advice to the Executive Branch.

CONCLUSION

1. A rescissory action allows a challenge to a ruling after all appeals have been exhausted but one of the parties believes the decision was seriously flawed.

2. I thank José Cordeiro de Araújo for drawing my attention to this point.

3. This interpretation was subject to judicial review before an out-of-court settlement put an end to litigation (see chapter 3).

4. Epigenetics is the study of heritable phenotype changes that do not involve alterations in the DNA sequence and pays particular attention to the influence of the environment on gene activity and expression. Genomics is concerned with the sequencing of the genome, ascribing functions to genes and understanding their structure. Postgenomics goes a step further, studying, for example, how genes are transcribed into messenger RNA and how they are expressed as proteins.

5. For an overview of the Roundup Ready patent portfolio, see Jefferson et al. (2016).

6. While proponents contend that gene editing is more precise than recombinant DNA techniques, critics argue that the GE process can nevertheless have on-target effects (unintended effects at the target site) and off-target effects (unintended effects in other parts of the genome).

7. I thank Maywa Montenegro for drawing my attention to this point.

REFERENCES

Abdelgawad, Walid. 2012. The Bt brinjal case: The first legal action against Monsanto and its Indian collaborators for biopiracy. *Biotechnology Law Report* 31(2): 136–9. https://doi.org/10.1089/blr.2012.9926.

ABSP (Agricultural Biotechnology Support Project). 2003. *Final technical report 1991–2003*. Michigan State University. https://pdf.usaid.gov/pdf_docs/Pdacd491.pdf.

ABSP-II (Agricultural Biotechnology Support Project II). 2005. Newsletter. South Asia 1(1) (September). On file with the author.

ABSP-II (Agricultural Biotechnology Support Project II). n.d. absp2.cornell.edu.

Agarwal, Pankhuri, and Swaraj Paul Barooah, eds. n.d. *SpicyIP*. www.spicyip.com.

Agha, Eram. 2018. Modi govt being blackmailed by MNCs on GM Crops, say RSS-linked farmers' unions. *News18*. January 20. https://www.news18.com/news/india/modi-govt-being-blackmailed-by-mncs-on-gm-crops-say-rss-linked-farmers-unions-1637129.html.

Andersen, Regine. 2008. *Governing agrobiodiversity: Plant genetics and developing countries*. Aldershot: Ashgate.

Andersen, Walter, and Shridhar D. Damle. 2019. *Messengers of Hindu nationalism. How the RSS reshaped India*. London: C. Hurst.

Andow, David A. 2010. Bt brinjal: The scope and adequacy of the GEAC environmental assessment. http://www.gmwatch.org/files/Andow_Report_Bt_Brinjal.pdf.

Aoki, Keith. 2008. *Seed wars: Controversies and cases on plant genetic resources and intellectual property*. Durham, NC: Carolina Academic Press.

Araújo, José Cordeiro. 2010. *A Lei de Proteção de Cultivares: Análise de sua formulação e conteúdo.* Brasília: Câmara dos Deputados, Edições Câmara.

Arya, Shishir, and Snehlata Shrivastav. 2015. Seeds of doubt: Monsanto never had Bt cotton patent. *Times of India.* June 8. https://timesofindia.indiatimes.com/india /seeds-of-doubt-monsanto-never-had-bt-cotton-patent/articleshow/47578304.cms.

Ávila, Charlene. 2015. Da expectativa de direitos da Monsanto no Brasil sobre os pedidos de patentes da "tecnologia" Intacta RR2: Onde está de fato a inovação? *PIDCC* IV (8): 85–134. http://pidcc.com.br/artigos/082015/05082015.pdf.

Barbosa, Denis Borges. 2014. *Dois estudos sobre os aspectos jurídicos do patenteamento da tecnologia Roundup Ready no Brasil. A questão da soja transgênica.* PIDCC III (7): 330–468. http://pidcc.com.br/artigos/072014/16082014.pdf.

Barry, Gerald F., et al. 2006. Glyphosate-tolerant 5-enolpyruvylshikimate-3-phosphate synthases. US Patent RE39247, filed July 18, 2003, issued August 22, 2006.

Barry, Gerald F., et al. 2007. 5-Enolpiruvilshiquimato-3-fosfato sintases tolerantes ao glifosato. Brazilian Patent PI 1100008-2, filed June 12, 1996, issued April 24, 2007.

Baumer, João. 2005. Soja transgênica volta à Justiça. *O Estado de São Paulo.* Economia e Negócios, B7, September 25.

Beauchamp, Christopher. 2011. The pure thoughts of Judge Hand: A historical note on the patenting of nature. *NYU Law:* 1–39.

Bera, Sayantan, and Shreeja Sen. 2016. Centre tells Delhi high court Bt cotton's resistance to pests has waned. *Live Mint.* January 29. https://www.livemint.com/Politics /oZHYGceXXVZB3lit9PytEN/Centre-tells-Delhi-high-court-Bt-cottons-resistance-to -pest.html.

Bhardwaj, Mayank. 2016. Exclusive: Monsanto pulls new GM cotton seed from India in protest. *Reuters.* August 24. https://www.reuters.com/article/us-india-monsanto -exclusive-idUSKCN10Z1OX.

Bhardwaj, Mayank, Rupam Jain, and Tom Lasseter. 2017. Seed giant Monsanto meets its match as Hindu nationalists assert power in Modi's India. *Reuters.* March 28. https://www.reuters.com/investigates/special-report/monsanto-india/.

Bhardwaj, Mayank, and Aditya Kaira. 2021. Bayer's Monsanto, India's NSL settle long-running GM cotton seed dispute. *Reuters.* March 26. https://www.reuters.com /world/india/exclusive-bayers-monsanto-indias-nsl-settle-long-running-intellectual -property-2021-03-26/.

Bhat, R.V., and M.N. Madhyastha. 2007. Preserving the heritage of Mattu gulla: A variety of brinjal unique to Udupi district. *Current Science* 93(7): 905–6.

Bianchini. n.d. Soja RR2 IPRO (Intacta). On file with the author.

Boldrin, Michele, and David Levine. 2008. *Against intellectual monopoly.* Cambridge: Cambridge University Press.

Bornstein, Erica, and Aradhana Sharma. 2016. The righteous and the rightful: The technomoral politics of NGOs, social movements, and the state in India. *American Ethnologist* 43(1): 76–90.

Borowiak, Craig. 2004. Farmers' rights: Intellectual property regimes and the struggle over seeds. *Politics & Society* 32(4): 511–43.

Boyle, James. 2008. *The public domain: Enclosing the commons of the mind.* New Haven, CT: Yale University Press.

Bragdon, Susan H. 2020. Global legal constraints: How the international system fails small-scale farmers and agricultural biodiversity, harming human and planetary health, and what to do about it. *American University International Law Review* 36(1): 1–50.

Brown, Symeon. 2018. Fake it till you make it: Meet the wolves of Instagram. *The Guardian.* April 19. https://www.theguardian.com/news/2018/apr/19/wolves-of-insta gram-jordan-belmont-social-media-traders.

Brush, Stephen B. 2013. Agrobiodiversity and the law: Regulating genetic resources, food security and cultural diversity. *Journal of Peasant Studies* 40(2): 447–67.

CAG (Comptroller and Auditor General of India). 2010. Environment Audit Report. Report no. 17 of 2010–11. New Delhi. https://cag.gov.in/cag_old/sites/default/files /audit_report_files/Union_Compliance_Scientific_Department_Environment_17 _2010.pdf.

Callahan, Gary. 2009. Comments regarding agriculture and antitrust enforcement issues restraints on competition in sales of off-patent agrochemicals. Letter to Legal Policy Division, Antitrust Division, US Dept of Justice. Via email. July 31. https:// www.justice.gov/atr/public/workshops/ag2010/015/AGW-14481-a.doc.

Calvert, Jane, and Pierre-Benoît Joly. 2011. How did the gene become a chemical compound? The ontology of the gene and the patenting of DNA. *Social Science Information* 50(2): 1–21.

Canadian Press. 2001. Monsanto tangles with more Canadian farmers on licensing. July 19. http://www.connectotel.com/gmfood/cp190701.txt.

Canfield, Matthew. Property regimes. 2020. In *The Oxford Handbook of Law and Anthropology,* ed. Marie-Claire Foblets, Mark Goodale, Maria Sapignoli, and Olaf Zenker. Advance online publication. https://doi.org/10.1093/oxfordhb/978019884 0534.013.23.

CBD (Convention on Biological Diversity). n.d. [1]. Brazil—Country profile. https:// www.cbd.int/countries/?country=br.

CBD (Convention on Biological Diversity). n.d. [2]. India—Country profile. https:// www.cbd.int/countries/?country=in.

CCI (Competition Commission of India). 2016. *Department of Agriculture v. MMB.* Reference Case no. 22 of 2015, and *Nuziveedu v. MMB,* Case no. 107 of 2015.

Center for Food Safety and Save Our Seeds. 2013. Brief as Amici Curiae, in support of petitioner. *Bowman v. Monsanto*, 569 U.S. 278.

CGPDTM (Controller General of Patents, Designs and Trade Marks). 2013. First examination report. May 23. On file with the author.

Charles, Daniel. 2001. *Lords of the harvest: Biotech, big money and the future of food*. Cambridge, MA: Perseus.

Choudhary, Bhagirath, and Kadambini Gaur. 2009. The development and regulation of Bt brinjal in India (eggplant/aubergine). ISAAA Briefs 38. Ithaca, NY: ISAAA.

Chowdhury, Nupur, and Nidhi Srivastava. 2010. Decision on Bt-brinjal: Legal issues. *Economic and Political Weekly* 45(15): 18–22.

Chowgule, Ashok. 2015. Govt must stand up to anti-science opponents of GM technology: Ashok Chowgule. *Smart Indian Agriculture*. https://www.smartindian agriculture.com/vhp-leader-wants-govt-to-take-on-anti-science-opponents-of-gm -technology/.

Claeys, Priscilla. 2015. *Human rights and the food sovereignty movement: Reclaiming control*. New York: Routledge.

Clapp, Jennifer. 2021. The problem with growing corporate concentration in the global food system. *Nature Food* 2: 404–8. https://doi.org/10.1038/s43016-021-00297-7.

ClicRBS. 2005. Justiça nega suspensão do pagamento de royalties à Monsanto. February 17. https://www.clicrbs.com.br/especial/rs/nossomundo/19,997,782490.

COAD. 2012. Royalties da Monsanto: Ação de sojicultores tem alcance nacional. https://coad.jusbrasil.com.br/noticias/3148934/royalties-da-monsanto-acao-de -sojicultores-tem-alcance-nacional.

Cohen, Marc J., and Anitha Ramanna. 2007. Public access to seeds and the human right to adequate food. In *Global obligations for the human right to food*, ed. G. Kent, 161–90. Lanham, MD: Rowman and Littlefield.

Consultor Jurídico. 2005. Produtor é isentado de pagar royalties da soja à Monsanto. January 11. http://www.conjur.com.br/2005-jan-11/produtor_isentado_pagar_royal ties_soja_monsanto.

Coombe, Rosemary J. 1998. Intellectual property, human rights and sovereignty: New dilemmas in international law posed by the recognition of indigenous knowledge and the conservation of biodiversity. *Indiana Journal of Global Legal Studies* 6(1): 59–115.

Coombe, Rosemary, and Susannah Chapman. 2020. Ethnographic explorations of intellectual property. In *Oxford Research Encyclopedia of Anthropology*. Online publication. https://doi.org/10.1093/acrefore/9780190854584.013.115.

Corbin, David R., and C. P. Romano. 2008. Methods for transforming plants to express bacillus thuringiensis delta-endotoxins. Indian patent 214436, filed May 1, 2001, issued February 12, 2008.

Correa, Carlos M., et al. 2015. *Plant variety protection in developing countries: A tool for designing sui generis plant variety protection system. An alternative to UPOV 1991.* APBREBES.

Costa, Pryscila. 2005. Produtor do RS fica livre de pagar royalties a Monsanto. *Consultor Jurídico.* April 15. http://www.conjur.com.br/2005-abr-15/produtor_rs_fica_livre _pagar_royalties_monsanto.

Cullet, Philippe. 2005a. Case law analysis. *Monsanto v Schmeiser: A landmark decision concerning farmer liability and transgenic contamination. Journal of Environmental Law* 17(1): 83–108.

Cullet, Philippe. 2005b. Seed regulation, food security and sustainable development. *Economic and Political Weekly* 40(32): 3607–13.

Cullet, Philippe. 2007. Human rights and intellectual property protection in the TRIPS era. *Human Rights Quarterly* 29(2): 403–29.

Dalrymple, William. 2019. *The anarchy: The relentless rise of the East India Company.* London: Bloomsbury.

Damodaran, Harish. 2016. Janus-faced policy to IPR: Farm vs drugs? *Indian Express.* March 10. https://indianexpress.com/article/india/india-news-india/bt-cotton -monsanto-janus-faced-policy-to-ipr-farm-vs-drugs/.

Das, Sandip. 2016. Bt cotton crop area falls for first time. *Financial Express.* October 22. https://www.financialexpress.com/market/commodities/bt-cotton-crop-area-falls -for-first-time/426413/.

de Alencar, Gisela, and Marco C. van der Ree. 1996. 1996: An important year for Brazilian biopolitics? *Biotechnology and Development Monitor* 27: 21–2.

de Schutter, Olivier. 2009. Seed policies and the right to food: Enhancing agrobiodiversity and encouraging innovation. United Nations General Assembly. A/64/170.

Deshpande, Vivek. 2016. Monsanto patent under cloud as Bt cotton prone to pink bollworm. *Indian Express.* March 8. https://indianexpress.com/article/india/india-news -india/monsanto-patent-under-cloud-as-bt-cotton-prone-to-pink-bollworm/.

Desmarais, A. A. 2007. *La Via Campesina: Globalization and the power of peasants.* Black Point, Nova Scotia: Fernwood.

Down to Earth. 2019. Pepsico India withdraws all cases against Gujarat potato farmers. May 10. https://www.downtoearth.org.in/news/agriculture/pepsico-india -withdraws-all-cases-against-gujarat-potato-farmers-64482.

Dutfield, Graham. 2006. Patent systems as regulatory institutions. *Indian Economic Journal* 54(1): 62–90.

Dutfield, Graham. 2008. Turning plant varieties into intellectual property: The UPOV Convention. In *The future control of food: A guide to international negotiations and rules on intellectual property, biodiversity and food security,* ed. Geoff Tansey and Tasmin Rajotte, 27–47. London: Earthscan and IDRC.

Dutfield, Graham. 2011. Food, biological diversity and intellectual property: The role of the International Union for the Protection of New Varieties of Plants (UPOV). Global Economic Issue Publications. Intellectual Property Issue Paper 9. Geneva: Quaker United Nations Office. https://quno.org/sites/default/files/resources /UPOV%2Bstudy%2Bby%2BQUNO_English.pdf.

Economic Times. 2015. Hybrid seed producers want government to move CCI against Mahyco Monsanto. October 15. https://economictimes.indiatimes.com/news /economy/agriculture/hybrid-seed-producers-want-government-to-move-cci-against -mahyco-monsanto/articleshow/49367624.cms.

Edelman, Marc, and Saturnino M. Borras. 2016. *Political Dynamics of Transnational Agrarian Movements*. Rugby, UK: Practical Action/Black Point, Nova Scotia: Fernwood.

Egelie, Knut J., Gregory D. Graff, Sabina P. Strand, and Berit Johansen. 2016. The emerging patent landscape of CRISPR-Cas gene editing technology. *Nature Biotechnology* 34(10): 1025–31.

EMBRAPA n.d. EMBRAPA Recursos Genéticos e Biotecnologia. https://www.embrapa .br/recursos-geneticos-e-biotecnologia.

ESG (Environment Support Group). 2010a. An enquiry into certain legal issues relating to the approval of Bt brinjal by the Genetic Engineering Approval Committee of the Union Ministry of Environment and Forests. Submission made to Shri. Jairam Ramesh in the Public Consultation on GEAC approval to Bt Brinjal. February 6. http://static.esgindia.org/campaigns/brinjal/action.html.

ESG (Environment Support Group). 2010b. Letter to KBB regarding violations of Biological Diversity Act, 2002 in matters relating to access and utilisation of local brinjal varieties for development of Bt Brinjal by M/s Mahyco and ors., and related issues. February 15. http://www.esgindia.org/sites/default/files/campaigns/brinjal /press/esg-karbioboard-btbrinjal-petition-12021.pdf.

ESG (Environment Support Group). 2013. Transferring prosecuting officers exposes Indian and Karnataka governments' weak intent to tackle biopiracy by Mahyco/ Monsanto and others. Press release. Bengaluru. March 9. https://esgindia.org/new /campaigns/transferring-prosecuting-officers-exposes-indian-and-karnataka-govern ments-weak-intent-to-tackle-biopiracy-by-mahyco-monsanto-and-others/.

ESG (Environment Support Group). n.d. http://esgindia.org.

ETC Group. 2007. Updated: The world's top 10 seed companies—2006. https://www .etcgroup.org/content/updated-worlds-top-10-seed-companies-2006.

Ewens, Lara E. 2000. Seed wars: Biotechnology, intellectual property and the quest for high yield seeds. *Boston College International & Comparative Law Review* 23: 285–310.

FAMATO. 2013. Análise jurídica: Acordo de licenciamento de tecnologia e quitação geral. https://www.noticiasagricolas.com.br/noticias/soja/116757-caso-monsanto -acordo-de-licenciamento-de-tecnologia-e-quitacao-geral--rebatido-pela-aprosoja .html.

FAO (Food and Agriculture Organization). 2001. International Treaty on Plant Genetic Resources for Food and Agriculture. Rome. November 3.

Federizzi, Luiz Carlos. 2011. Parecer técnico. Processo no. 001/1090106915–2. On file with the author.

Fernandes, Vivian. 2017. PPVFR Act dispensing with NOCs: Agriculture ministry rendered hollow the plant trait patents of companies like Monsanto. *Financial Express.* June 21. https://www.financialexpress.com/economy/ppvfr-act-dispensing-with-nocs -agriculture-ministry-rendered-hollow-the-plant-trait-patents-of-companies-like -monsanto/728559/.

Fernandes, Vivian. 2018. Bt cotton seed price control: How seed companies shot themselves in the foot. *Financial Express.* March 15. https://www.financialexpress .com/opinion/bt-cotton-seed-price-control-how-seed-companies-shot-themselves-in -the-foot/1099219/.

FETAG. 2009. FETAG questiona na Justiça pagamento de royalties. June 12. http:// fetagrs.org.br/fetag-questiona-na-justia-pagamento-de-royalties/.

Filomeno, Felipe Amin. 2014. *Monsanto and intellectual property in South America.* Basingstoke, UK: Palgrave Macmillan.

Fincher, Karen L. 2012. Sequência de DNA de promotor quimérico, constructos de DNA, método de expressar uma sequência de DNA estrutural em uma planta e método de controlar ervas daninhas. PI 0016460–7, filed December 12, 2000, issued October 2, 2012.

Flachs, Andrew. 2019. *Cultivating knowledge: Biotechnology, sustainability, and the human cost of cotton capitalism in India.* Tucson: University of Arizona Press.

FOEI (Friends of the Earth International). 2006. Who benefits from GM crops? Monsanto and the corporate-driven genetically modified crop revolution. Issue 110. Amsterdam: Friends of the Earth International. https://friendsoftheearth.eu/wp -content/uploads/2012/06/who_benefits_from_gm_crops_jan_2006.pdf.

Folha do Cerrado. 2014. Soja Intacta: Monsanto consegue patente e deverá estampar selo de propriedade até 2022. March 12. On file with the author.

Fowler, Cary, and Pat Mooney. 1990. *Shattering: Food, politics and the loss of genetic diversity.* Tucson: University of Arizona Press.

Friedmann, Harriet, and Philip McMichael. 1989. Agriculture and the state system: The rise and decline of national agricultures. *Sociologia Ruralis* 29(2): 93–117.

GAIA/GRAIN. 1998. Ten reasons not to join UPOV: Global trade and biodiversity in conflict. Issue 2. https://grain.org/e/1.

Ganesan, Arumugamangalam Venkatachalam. 2015. Negotiating for India. In *The making of the TRIPS Agreement: Personal insights from the Uruguay Round negotiations,* ed. Jayashree Watal and Anthony Taubman, 211–38. Geneva: World Trade Organization.

GATT. 1987. Suggestion by the United States for achieving the negotiating objective. MTN.GNG/NG11/W/14.

GATT. 1988. Guidelines and objectives proposed by the European community for the negotiations on trade related aspects of substantive standards of intellectual property rights. MTN.GNG/NG11/W/26.

GATT. 1989a. Communication from India. Negotiating group on trade-related aspects of intellectual property rights, including in counterfeit goods. MTN.GNG/NG11/W/37.

GATT. 1989b. Communication from Brazil: Negotiating group on trade-related aspects of intellectual property rights, including in counterfeit goods. MTN.GNG/NG11/W/57.

GATT. 1990. Communication from Argentina, Brazil, Chile, China, Colombia, Cuba, Egypt, India, Nigeria, Peru, Tanzania, and Uruguay. MTN.GNG/NG11/W/71.

Gene Campaign. 2003. Oppose UPOV! Save farmers! Gene Campaign's legal action against Indian government. New Delhi. On file with the author.

Glover, Dominic. 2009. Undying promise: Agricultural biotechnology's pro-poor narrative, ten years on. STEPS Working Paper 15. Brighton, UK: STEPS Centre.

Glover, Dominic. 2010a. The corporate shaping of GM crops as a technology for the poor. *Journal of Peasant Studies* 37(1): 67–90.

Glover, Dominic. 2010b. Is Bt cotton a pro-poor technology? A review and critique of the empirical record. *Journal of Agrarian Change* 10(4): 482–509.

GoI (Government of India). 1970. Patents Act 1970, as amended by Patents (Amendments) Act 2005.

GoI (Government of India). 2001. Protection of Plant Varieties and Farmers' Rights Act.

GoI (Government of India). 2002a. Biological Diversity Act.

GoI (Government of India). 2002b. Competition Act.

GoI (Government of India). 2015. Cotton Seeds Price (Control) Order.

GoI (Government of India). 2020. All India Report on Agriculture Census 2015–16. Department of Agriculture, Cooperation & Farmers Welfare, Ministry of Agriculture and Farmers Welfare. http://agcensus.nic.in/document/agcen1516/ac_1516_report_final-220221.pdf.

Golay, Christophe. 2017. The right to seeds and intellectual property rights. Research brief. Geneva: Geneva Academy of International Humanitarian and Human Rights. https://www.ohchr.org/Documents/HRBodies/HRCouncil/WGPleasants/Session4/Geneva_Academy-Right_to_seeds.pdf.

GRAIN. 1996. UPOV: Getting a free TRIPs ride? https://grain.org/e/321.

GRAIN. 2006. Andhra Pradesh files case against Bt cotton royalty. January 3. https://www.grain.org/e/2186.

GRAIN and LVC. 2015. Seed laws that criminalise farmers: Resistance and fightback. https://www.grain.org/e/5142.

Gutiérrez Escobar, Laura, and Elizabeth Fitting. 2016. The *Red de Semillas Libres*: Contesting biohegemony in Colombia. *Journal of Agrarian Change* 16(4): 711–9.

Halewood, Michael, Isabel López Noriega, and Selim Louafi. 2013. *Crop genetic resources as a global commons: Challenges in international law and governance*. New York: Routledge.

Haugen, Hans Morten. 2007. Patent rights and human rights: Exploring the relationships. *Journal of World Intellectual Property* 10(2): 97–124.

Haugen, Hans Morten. 2020. The UN Declaration on Peasants' Rights (UNDROP): Is Article 19 on seed rights adequately balancing intellectual property rights and the right to food? *Journal of World Intellectual Property* 23(3–4): 288–309.

Helfer, Laurence R. 2004. Intellectual property rights in plant varieties. International legal regimes and policy options for national governments. FAO legislative study 85. Rome: FAO.

Helfer, Laurence. 2018. Intellectual property and human rights: Mapping an evolving and contested relationship. In *The Oxford Handbook of Intellectual Property Law*, ed. Rochelle Dreyfus and Justine Pila. https://doi.org/10.1093/oxfordhb/9780198758457.013.9.

Helfer, Laurence R., and Graeme W. Austin. 2011. *Human rights and intellectual property: Mapping the global interface*. New York: Cambridge University Press.

Herring, Ronald J. 2007. Stealth seeds: Bioproperty, biosafety, biopolitics. *Journal of Development Studies* 43(1): 130–57.

Hindu. 2016. Bt cotton row: Monsanto threatens to re-evaluate India biz. March 4. http://www.thehindu.com/business/bt-cotton-row-monsanto-threatens-to-reevaluate-india-biz/article8313841.ece.

Howard, Philip H. 2015. Intellectual property and consolidation in the seed industry. *Crop Science* 55: 1–7.

Howard, Philip H. 2018. Global seed industry changes since 2013. https://philhoward.net/2018/12/31/global-seed-industry-changes-since-2013/.

Huntington, Samuel P. 1996. *The clash of civilizations and the remaking of world order*. New York: Simon and Schuster.

IAASTD (International Assessment of Agricultural Knowledge, Science and Technology for Development). 2009. Executive summary of the synthesis report.

IPES-Food. 2017. Too big to feed: Exploring the impacts of mega-mergers, consolidation and concentration of power in the agri-food sector. http://www.ipes-food.org /_img/upload/files/Concentration_FullReport.pdf.

ISAAA (International Service for the Acquisition of Agri-Biotech Applications). 2019. Global status of commercialized biotech/GM crops in 2019. ISAAA Brief 55. Ithaca, NY: ISAAA.

ISAAA (International Service for the Acquisition of Agri-Biotech Applications). n.d. GM approval database. GM crop events list. Bt Brinjal EE1. http://www.isaaa.org /gmapprovaldatabase/event/default.asp?EventID=351.

Jadhav, Radheshyam. 2019. HTBT cotton widely cultivated despite ban, finds Central team. *The Hindu.* June 12. https://www.thehindubusinessline.com/economy /agri-business/htbt-cotton-widely-cultivated-despite-ban-finds-central-team /article27891145.ece.

Jaffe, Adam B., and Josh Lerner. 2007. *Innovation and its discontents: How our broken patent system is endangering innovation and progress, and what to do about it.* Princeton, NJ: Princeton University Press.

Jayaraman, Killugudi S. 2012. India investigates Bt cotton claims. *Nature.* February 14. https://www.nature.com/news/india-investigates-bt-cotton-claims-1.10015.

Jefferson, David J., Gregory D. Graff, Cecilia L. Chi-Ham, and Alan B. Bennett. 2016. The emergence of agbiogenerics. *Nature Biotechnology* 33(8): 819–23.

Jefferson, David J., and Meenu S. Padmanabhan. 2016. Recent evolutions in intellectual property frameworks for agricultural biotechnology: A worldwide survey. *Asian Biotechnology and Development Review* 18(1): 17–37.

Jishnu, Latha. 2010a. Not a nice trait to have: Monsanto is back in the courts on the issue of royalty or trait fees it charges for its genetically modified Bt cotton. *Business Standard.* April 1. http://www.business-standard.com/article/opinion/latha-jishnu-not -a-nice-trait-to-have-110040100057_1.html.

Jishnu, Latha. 2010b. Battle royal over Bt cotton. *Business Standard.* May 28. https:// www.business-standard.com/article/economy-policy/battle-royal-over-bt-cotton -royalty-110052800037_1.html.

Jishnu, Latha. 2010c. An odd royalty calculus. *Business Standard.* June 24. https:// www.business-standard.com/article/opinion/latha-jishnu-an-odd-royalty-calculus -110062400010_1.html.

Jishnu, Latha. 2015. Sangh Parivar groups fighting GM mustard run into a new opponent—Sangh Parivar groups. *Scroll In.* December 7. https://scroll.in/article/772215 /sangh-parivar-groups-fighting-gm-mustard-run-into-a-new-opponent-sangh-parivar -groups.

Jurrens, Damion. 2018. MON/BAYN: No decision yet in Brazilian patent fight, Court awaits Monsanto response. January 24. https://reorg.com/mon-bayn-no-decision -yet-in-brazilian-patent-fight-court-awaits-monsanto-response/.

Kang, Bhavdeep. 2016. RSS stand on Bt cotton forced government's hand on Monsanto. *The Wire*. March 13. https://thewire.in/agriculture/rss-stand-on-bt-cotton -forced-governments-hand-on-monsanto.

Kassai, Lucia. 2005. Monsanto cobra desde já royalties da soja. *Gazeta Mercantil*. August 23. https://www.agrolink.com.br/noticias/monsanto-cobra-desde-ja-royalties -da-soja_30739.html.

Kaveri Seeds. 2015. Temporary thread for clarifications. http://forum.valuepickr .com/t/kaveri-seeds-temporary-thread-for-clarifications/2956.

KBB (Karnataka Biodiversity Board). 2010. Forest, Ecology and Environment Dept. no. KBB/BT/77/09–10/2004. Bengaluru. March 10. Petition for Special Leave to Appeal no. 7951/2014. Annexure P-7, 136–43. On file with the author.

KBB (Karnataka Biodiversity Board). 2011. Forest, Ecology and Environment Dept. no. KBB/Bt.B/71/11–12/42. Bengaluru. May 28. Petition for Special Leave to Appeal no. 7951/2014. Annexure P-9, 146–49. On file with the author.

KBB (Karnataka Biodiversity Board). 2012. Proceedings of the 19th board meeting. Bengaluru. January 20. Petition for Special Leave to Appeal no. 7951/2014. Annexure P-14, 213–29. On file with the author.

Kent, Lawrence. 2007. What's the holdup? Addressing constraints to the use of plant biotechnology in developing countries. *AgBioForum* 7(1–2): 63–9.

Kloppenburg, Jack. 2004 [1988]. *First the seed: The political economy of plant biotechnology, 1492–2000*. Madison: University of Wisconsin Press.

Kochupillai, Mrinalini. 2016. *Promoting sustainable innovations in plant varieties*. Berlin: Springer.

Kock, Michael A. 2021. Open intellectual property models for plant innovations in the context of new breeding technologies. *Agronomy* 11(6). https://doi.org/10.3390 /agronomy11061218.

Kranthi, Keshav Raj. 2012. Bt cotton: Questions and answers. Mumbai: Indian Society for Cotton Improvement.

Kranthi, Keshav Raj. 2016. Technology and agriculture: Messed in India! *Indian Express*. July 3. https://indianexpress.com/article/india/india-news-india/technology -and-agriculture-messed-in-india/.

Krimsky, Sheldon. 2019. *GMOs decoded: A skeptic's view of genetically modified food*. Boston: MIT Press.

Krishnakumar, Asha. 2004. Bt cotton, again. *Frontline* 21(10). May 8–21. https:// frontline.thehindu.com/science-and-technology/article30222477.ece.

Kurmanath, K.V. 2010. AP fixes royalty for Monsanto cotton seed. *The Hindu Business Line*. May 4. https://www.thehindubusinessline.com/todays-paper/tp-ag ri-biz-and-commodity/AP-fixes-royalty-for-Monsanto-cotton-seed/article20039 712.ece.

Kurmanath, K.V. 2016. Why should we not revoke Bollgard II patent, Centre asks Mahyco Monsanto. *The Hindu Business Line*. March 10. https://www.thehindubusi nessline.com/economy/agri-business/why-should-we-not-revoke-bollgard-ii-patent -centre-asks-mahyco-monsanto/article8337510.ece.

Kurmanath, K.V. 2021. "Seed factions" pact will clear agri-tech logjam. *The Hindu Business Line*. July 12. https://www.thehindubusinessline.com/economy/agri-business /seed-factions-pact-will-clear-agri-tech-logjam/article35288588.ece.

Kuyek, Devlin, et al. 2000. ISAAA in Asia: Promoting corporate profits in the name of the poor. *GRAIN Report*. October 25. https://www.grain.org/e/35.

Lakshmikumaran & Sridharan (law firm). 2014a. Re: Indian patent application no. 368/ DEL/2006, dated February 10, 2006. May 6. On file with the author.

Lakshmikumaran & Sridharan (law firm). 2014b. Re: Indian application no. 368/ DEL/2006, filed on February 10, 2006. July 2. On file with the author.

Lapegna, Pablo, and Tamara Perelmuter. 2020. Genetically modified crops and seed/ food sovereignty in Argentina: Scales and states in the contemporary food regime. *Journal of Peasant Studies* 47(4): 700–19.

Lock, Margaret. 2005. Eclipse of the gene and the return of divination. *Current Anthropology* 46: S47–70.

Lok Sabha. 2012a. Performance of the Ministry of Environment and Forests. Fifty-seventh report. Public Accounts Committee (2011–2012). Fifteenth Lok Sabha. New Delhi: Lok Sabha Secretariat.

Lok Sabha. 2012b. Ministry of Agriculture (Department of Agriculture and Coopera-tion). Cultivation of genetically modified food crops: Prospects and effects. Thirty-seventh report. Committee on Agriculture (2011–2012). Fifteenth Lok Sabha. New Delhi: Lok Sabha Secretariat.

Lunardi, Soraya, and Dimitri Dimoulis. 2017. O custo social da inércia do STF: Réquiem da ADI 4.234. July 26. https://jota.info/colunas/supra/o-custo-social-da -inercia-do-stf-requiem-da-adi-4-234-26072017.

Mahler, Anne Garland. 2017. Global South. In *Oxford bibliographies in literary and critical theory*, ed. Eugene O'Brien. New York: Oxford University Press.

Mahyco. 2006. Transgenic brinjal (Solanum Melongena) comprising EE-1 event. Patent application no. 368/DEL/2006, filed February 10, 2006. On file with the author.

Mahyco. 2013. Statements of objections/Counter affidavit on behalf of the respon-dent no. 6. In the High Court of Karnataka at Bangalore. WP 41532/2012. June 15. On file with the author.

Mahyco, Sathguru, and UAS Dharwad (Maharashtra Hybrid Seed Co. Ltd., Sathguru Management Consultants Private Limited, and University of Agricultural Sciences

Dharwad). 2005. Sublicense agreement. April 2. http://esgindia.org/sites/default/files /campaigns/brinjal/press/d-uas-dharwad-mahyco-agreement-2005-1-15.pdf.

Mahyco and TNAU (Maharashtra Hybrid Seeds Company Limited and Tamil Nadu Agricultural University). 2005. Material transfer agreement. March 20. http://www .esgindia.org/sites/default/files/campaigns/brinjal/press/e-bt-brinjal-mta-tnau -mahyco-mar-2005000.pdf.

Manjunatha, B.L., D. U. M. Rao, M. B. Dastagiri, J. P. Sharma, and R. Roy Burman. 2015. Need for government intervention in regulating seed sale price and trait fee: A case of Bt cotton. *Journal of Intellectual Property Rights* 20: 375–87.

Marden, Emily, R. Nelson Godfrey, and Rachael Manion. 2016. *The intellectual property–regulatory complex: Overcoming barriers to innovation in agricultural genomics.* Vancouver: University of British Columbia Press.

Massarini, Luisa. 2012. Monsanto may lose GM soya royalties throughout Brazil. *Nature News.* June 15. https://doi.org/10.1038/nature.2012.10837.

Matthews, Duncan. 2002. *Globalizing intellectual property rights: The TRIPS agreement.* London: Routledge.

McAfee, Kathleen. 2003. Neoliberalism on the molecular scale: Economic and genetic reductionism in biotechnology battles. *Geoforum* 34: 203–19.

McCann, Michael. 2004. Law and social movements. In *The Blackwell Companion to Law and Society*, ed. Austin Sarat, 506–22. Oxford: Blackwell.

McMichael, Philip. 2009. A food regime genealogy. *Journal of Peasant Studies* 36(1): 139–69.

McMichael, Philip. 2013. *Food regimes and agrarian questions.* Halifax: Fernwood.

MDA (Ministério do Desenvolvimento Agrário). 2008. Mais alimentos: Um plano da agricultura familiar para o Brasil. Plano Safra da Agricultura Familiar 2008/09. Brasília: MDA.

Mehta, Pradeep S. 2006. Of virus, seeds, patent, competition. *The Hindu Business Line.* November 17. https://cuts-ccier.org/of-virus-seeds-patents-competition/.

Ministry of Commerce and Industry. 2002. India and the WTO. A monthly newsletter of the Ministry of Commerce and Industry 4(5) (May). http://commerce.nic.in /writereaddata/publications/wto_may2002.pdf.

MoEF (Minister of Environment and Forests). 2006. Notification S.O.1911(E). Published in *Gazette of India*, Extraordinary, Part II, Section 3, Sub-section (ii). November 8.

MoEF (Minister of Environment and Forests). 2009. Notification S.O.2726(E). Published in *Gazette of India*, Extraordinary, Part II, Section 3, Sub-section (ii). October 26.

MoEF (Ministry of Environment and Forests). 2010a. Decision on commercialisation of Bt-Brinjal. February 9.

MoEF (Minister of Environment and Forests). 2010b. Clarification on MoEF Notification of October 26, 2009, on biological resources notified as normally traded commodities. February 16.

MoEF (Ministry of Environment, Forests and Climate Change). 2014. Notification S.O. 3232(E). Published in *Gazette of India*, Extraordinary, Part II, Section 3, Subsection (ii). December 14.

Mohan, Vishwa. 2016. RSS-linked group brings all anti-GM NGOs together on one platform to oppose transgenic mustard. *Times of India*. September 30. https://timesofindia.indiatimes.com/india/rss-linked-group-brings-all-anti-gm-ngos-together-on-one-platform-to-oppose-transgenic-mustard/articleshow/54612728.cms.

Monsanto. 2003. Comunicado da Monsanto para sojicultores. *Correio do Povo*. September 16. On file with the author.

Monsanto. 2005. Campanha "Tecnologia Roundup Ready. Você sabe o valor que ela tem." *Correio do Povo*. February 5. On file with the author.

Monsanto. 2010. Pink bollworm resistance to GM cotton in India. Press release. On file with the author.

Monsanto. n.d. Acordo geral para licenciamento de direitos de propriedade intelectual da tecnologia Roundup Ready®. On file with the author.

Montenegro de Wit, Maywa. 2020. Democratizing CRISPR? Stories, practices and politics of science and governance on the agricultural gene editing frontier. *Elementa Science of the Anthropocene* 8(9). http://doi.org/10.1525/elementa.405.

Motta, Renata. 2016. *Social mobilization, global capitalism and struggles over food.* London: Routledge.

Moudgil, Manu. 2017. Every seed makes a political statement. *YourStory*. https://yourstory.com/2017/06/seed-economy.

Müller, Birgit. 2006. Infringing and trespassing plants: Patented seeds at dispute in Canada's courts. *Focaal* 48: 83–98.

Muniz, Mariana. 2018. Após nove anos, STF pauta julgamento de ação sobre patentes de remédios. *JOTA*. July 16. https://www.jota.info/justica/apos-nove-anos-stf-pauta-julgamento-de-acao-sobre-patentes-de-remedios-16072018.

NAS (National Academies of Sciences, Engineering, and Medicine). 2016. *Genetically engineered crops: Experiences and prospects.* Washington, DC: National Academies Press.

NBA (National Biodiversity Authority). 2011a. Proceedings of the 20th Authority Meeting. New Delhi. June 20. Petition for Special Leave to appeal no. 7951/2014. Annexure P-10, 150–68. On file with the author.

NBA (National Biodiversity Authority). 2011b. Proceedings of the 22nd meeting of NBA. Chennai. November 22. Petition for Special Leave to appeal no. 7951/2014. Annexure P-13, 194–212. On file with the author.

NBA (National Biodiversity Authority). 2012. Proceedings of the 23rd meeting of NBA. Chennai. February 28. Petition for Special Leave to appeal no. 7951/2014. Annexure P-15, 230–46. On file with the author.

NBA (National Biodiversity Authority). 2013. Statements of objections filed by the first respondent. In the High Court of Karnataka at Bangalore. WP 41532/2012. Annexure R2. On file with the author.

NBA (National Biodiversity Authority). 2014. NBA/TechAppl/9/607/13/14–15/275. May 2. On file with the author.

NBPGR (National Bureau of Plant Genetic Resources). n.d. http://www.nbpgr.ernet .in/Divisions_and_Units/Conservation.aspx.

Newell, Peter. 2006. Corporate power and "bounded autonomy" in the global politics of biotechnology. In *The international politics of genetically modified food*, ed. Robert Falkner, 67–84. London: Palgrave Macmillan.

Newell, Peter. 2007. Biotech firms, biotech politics: Negotiating GMOs in India. *Journal of Environment and Development* 16: 183–206.

Newell, Peter. 2008. Trade and biotechnology in Latin America: Democratization, contestation and the politics of mobilization. *Journal of Agrarian Change* 8 (2–3): 345–76.

NRC (National Research Council). 1990. *Plant biotechnology research for developing countries*. Report of a panel of the Board on Science and Technology for International Development. Washington, DC: National Academies Press.

OCGPDT (Office of Controller General of Patents, Designs & Trademarks). 2008. *Manual of Patent Office Practice and Procedure*. Mumbai: OCGPDT.

OCGPDT (Office of Controller General of Patents, Designs & Trademarks). 2013. *Guidelines for Examination of Biotechnology Applications for Patent*. Mumbai: OCGPDT.

Official Gazette. 2001. United States Patent and Trademark Office 1247(3). June 19.

Oh, Cecilia. 2000. TRIPS and pharmaceuticals: A case of corporate profits over public health. *Third World Network*. http://www.twn.my/title/twr120a.htm.

PANAP (Pesticide Action Network Asia and the Pacific). 2012. *India's Bt brinjal battle*. Penang, Malaysia: PANAP.

Pantulu, C. Chitti. 2006. Seven states take on Monsanto. *DNA India*. June 9. http:// www.dnaindia.com/india/report-seven-states-take-on-monsanto-1034534.

Parayil, Govindan. 2003. Mapping technological trajectories of the Green Revolution and the Gene Revolution from modernization to globalization. *Research Policy* 32: 971–90.

Park, Chan, and Arjun Jayadev. 2011. Access to medicines in India: A review of recent concerns. In *Access to knowledge in India*, ed. Ramesh Subramanian and Lea Shaver, 78–108. London: Bloomsbury Academic.

Parthasarathy, Shobita. 2017 *Patent politics: Life forms, markets, and the public interest in the United States and Europe.* Chicago: University of Chicago Press.

Paschoal, Adilson Dias. 1986. Prefácio do tradutor. In *O escândalo das sementes: O domínio na produção de alimentos*, Pat. Roy Mooney, xiii–xxvi. São Paulo: Nobel.

Peavey, Tabetha Marie. 2014. Bowman v. Monsanto: Bowman, the producer and the end user. *Annual Review of Law and Technology* 29: 465–92.

Pechlaner, Gabriela. 2012. *Corporate crops: Biotechnology, agriculture and the struggle for control.* Austin: University of Texas Press.

Pechlaner, Gabriela, and Gerardo Otero. 2008. The third food regime: Neoliberal globalism and agricultural biotechnology in North America. *Sociologia Ruralis* 48(4): 351–71.

Pelaez, Victor, and Wilson Schmidt. 2000. A difusão dos OGM no Brasil: Imposições e resistências. *Estudos, Sociedade e Agricultura* 14: 5–31.

Pelaez, Victor, and Wilson Schmidt. 2004. Social struggles and the regulation of transgenic crops in Brazil. In *Agribusiness and society: Corporate responses to environmentalism, market opportunities and public regulation*, ed. Kees Jansen and Sietze Vellema, 232–60. London: Zed Books.

Peschard, Karine. 2010. Biological dispossession: An ethnography of resistance to transgenic seeds among small farmers in Southern Brazil. PhD diss., McGill University.

Peschard, Karine. 2014. Farmers' rights and food sovereignty: Critical insights from India. *Journal of Peasant Studies* 41(6): 1085–108.

Peschard, Karine. 2017. Seed wars and farmers' rights: Comparative perspectives from Brazil and India. *Journal of Peasant Studies* 44(1): 144–68.

Peschard, Karine, and Shalini Randeria. 2020. "Keeping seeds in our hands": The rise of seed activism. *Journal of Peasant Studies* 47(4): 613–47.

Pollack, Malla. 2004. Originalism, J.E.M., and the food supply, or will the real decision maker please stand up. *Journal of Environmental Law and Litigation* 19: 495–534.

PPVFR Authority (Protection of Plant Varieties and Farmers' Rights Authority). 2016. List of applications pending due to legal issues. Application status up to June 10, 2016. On file with the author.

PPVFR Authority. 2019. List of certificates issued up to February 28, 2019. On file with the author.

Pray, Carl E., and Latha Nagarajan. 2010. Price controls and biotechnology innovation: Are state government policies reducing research and innovation by the ag biotech industry in India? *AgBioForum* 13(4): 197–307.

Press Information Bureau. 2011. Funding of NBA on BT BRINJAL. Press release. Ministry of Environment and Forests. New Delhi. September 6.

Purdue, Derrick A. 2000. *Anti-GenetiX: The emergence of the anti-GM movement.* Aldershot: Ashgate.

RAFI. 1993. Control of cotton: The patenting of transgenic cotton. RAFI communiqué. July–August. http://www.etcgroup.org/fr/node/498.

RAFI. 1997. World's top 10 seed corporations. https://www.etcgroup.org/content/worlds-top-10-seed-corporations.

Ramakrishna, S. Monopoly skins: AP's fight against Monsanto. *Down to Earth.* January 31. https://www.downtoearth.org.in/news/monopoly-skins-7213.

Ramanjaneyulu, GV. 2016. Bt cotton seed prices and royalties—Issues of concern. February 14. http://ramoo.in/?p=288.

Rana, Muhammad Hasan. 2021. When seed becomes capital: Commercialization of Bt cotton in Pakistan. *Journal of Agrarian Change.* Advance online publication. https://doi.org/10.1111/joac.12422.

Randeria, Shalini. 2003a. Domesticating neo-liberal discipline: Transnationalisation of law, fractured states and legal plurality in the South. In *Entangled histories and negotiated universals,* ed. Wolf Lepenies, 146–82. Frankfurt: Campus Verlag.

Randeria, Shalini. 2003b. Cunning states and unaccountable international institutions: Legal plurality, social movements and rights of local communities to common property resources. *European Journal of Sociology* 44(1): 27–60.

Randeria, Shalini. 2007. The state of globalization: Legal plurality, overlapping sovereignties and ambiguous alliances between civil society and the cunning state in India. *Theory, Culture & Society* 24(1): 1–33.

Randeria, Shalini, and Ciara Grunder. 2011. Policy-making in the shadow of the World Bank: Resettlement and urban infrastructure in the MUTP (India). In *Policy worlds: Anthropology and the analysis of contemporary power,* ed. Chris Shore, Sue Wright, and Davide Però, 187–204. New York: Berghahn.

Rao, Chavali Kameswara. 2013. Charges of "biopiracy" and violation of provisions of the Indian biodiversity act against the developers of *Bt* brinjal. Bangalore: Foundation for Biotechnology Awareness and Education. http://www.plantbiotechnology.org.in/issue47.html.

Ravi, Bhavishyavani. 2013. Gene patents in India: Gauging policy by an analysis of the grants made by the Indian Patent Office. *Journal of Intellectual Property Rights* 18: 323–9.

Reddy, Prashant. 2012. NBA set to prosecute Monsanto's Indian subsidary: What about Cornell, USAID & the DBT? *SpicyIP.* October 4. https://spicyip.com/2012/10/nba-set-to-prosecute-monsantos-indian.html.

Reddy, Prashant. 2018a. Delhi High Court's judgment in *Monsanto v. Nuziveedu* delivers a deadly blow to the agro-biotech industry. *SpicyIP*. April 15. https://spicyip.com/2018/04/delhi-high-courts-judgment-in-monsanto-v-nuziveedu-delivers-a-deadly-blow-to-the-agro-biotech-industry.html.

Reddy, Prashant. 2018b. Can Monsanto's invention be protected as a plant variety and can it seek benefit-sharing from Nuziveedu? *SpicyIP*. May 15. https://spicyip.com/2018/05/can-monsantos-invention-be-protected-as-a-plant-variety-and-can-it-seek-benefit-sharing-from-nuziveedu.html.

Reis, Maria Rita. 2005. Propriedade intelectual, sementes e o sistema de cobrança de royalties implementado pela Monsanto no Brasil. On file with the author.

RFB (República Federativa do Brasil). 1996. Law no. 9,279, of May 14, 1996.

RFB (República Federativa do Brasil). 1997. Law no. 9,456, of April 28, 1997.

RFB (República Federativa do Brasil). 2003. Medida Provisória no. 131, of September 25. Estabelece normas para a comercialização da produção de soja da safra de 2004 e dá outras providências. Convertida pela Lei no. 10.814, de 15 de dezembro de 2003.

Riordan, Teresa. 1994. U.S. revokes cotton patents after outcry from industry. *New York Times*. December 8. https://www.nytimes.com/1994/12/08/business/us-revokes-cotton-patents-after-outcry-from-industry.html.

Rodrigues, Roberta L., Celso L.S. Lage, and Alexandre G. Vasconcellos. 2011. Intellectual property rights related to the genetically modified glyphosate soybeans in Brazil. *Anais da Academia Brasileira de Ciências* 83(2): 719–30.

Roychowdhury, Anumita. 1994. Revoked! *Down to Earth*. March 31. https://www.downtoearth.org.in/news/revoked-29643.

Sahai, Suman. 2002. India: Plant variety protection and farmers' rights legislation. In *Global intellectual property rights: Knowledge, access and development*, ed. Peter Drahos and Ruth Mayne, 214–23. Basingstoke: Palgrave Macmillan.

Saldanha, Leo, and Bhargavi Rao. 2011. Monsanto's brinjal biopiracy: A shocking exposé of callous disregard for biodiversity laws in India. *India Law News* 2(4): 26–8.

Sally, Madhvi, and Karunjit Singh. 2019. Monsanto abused dominant position in India: CCI probe. *Economic Times*. May 22. https://economictimes.indiatimes.com/news/economy/agriculture/monsanto-abused-dominant-position-in-india-cci-probe/articleshow/69437310.cms.

Santilli, Juliana. 2012. *Agrobiodiversity and the law: Regulating genetic resources, food security and cultural diversity.* New York: Routledge.

Sathguru. 2013. Counter affidavit on behalf of the respondent no. 8. In the High Court of Karnataka at Bangalore. WP 41532/2012. On file with the author.

Sathyarajan, Sachin, and Balakrishna Pisupati. 2017. Genetic modification technology deployment: Lessons from India. Forum for Law, Environment, Development and Governance. Chennai: FLEDGE.

Schapiro, Mark. 2018. *Seeds of resistance: The fights to save our food supply*. New York: Hot Books.

Scoones, Ian. 2006. *Science, agriculture and the politics of policy: The case of biotechnology in India*. New Delhi: Orient Longman.

Scoones, Ian. 2008. Mobilizing against GM crops in India, South Africa and Brazil. *Journal of Agrarian Change* 8(2–3): 315–44.

Sease, Edmund J., and Robert A. Hodgson. 2006. Plants are properly patentable under prevailing U.S. law and this is good public policy. *Drake Journal of Agricultural Law* 11: 327–51.

Sehgal, Rashme. 2015. Thanks to Modi's push, GM mustard set to hit Indian markets. *Rediff*. June 22. http://www.rediff.com/business/report/thanks-to-modis-push-gm-mustard-set-to-hit-indian-markets/20150622.htm.

Sell, Susan K. 2003. *Private power, public law: The globalization of intellectual property rights*. Cambridge: Cambridge University Press.

Sell, Susan K. 2009. Corporations, seeds and intellectual property rights governance. In *Corporate power in global agrifood governance*, ed. Jennifer Clapp and Doris Fuchs, 186–223. Cambridge, MA: MIT Press.

Seshia, Shaila. 2002. Plant variety protection and farmers' rights: Law-making and cultivation of varietal control. *Economic and Political Weekly* 37: 2741–7.

Shah, Esha. 2011. "Science" in the risk politics of Bt brinjal. *Economic and Political Weekly* XLVI: 31–8.

Shappley, Zachary W., et al. 2009. Cotton Event MON 15985 and compositions and methods for detection. Indian patent 232681, filed December 8, 2003, issued March 20, 2009.

Shashikant, Sangeeta, and François Meienberg. 2015. International contradictions on farmers' rights: The interrelations between the International Treaty, its Article 9 on farmers' rights, and relevant instruments of UPOV and WIPO. Third World Network and the Berne Declaration.

Shiva, Vandana. 2016a. Monsanto vs Indian farmers. *Medium*. March 27. https://medium.com/@drvandanashiva/monsanto-vs-indian-farmers-60b70d6760f3.

Shiva, Vandana. 2016b. Clash of civilisations: India's ancient love for nature is losing out to modern disregard for it. *Scroll In*. https://scroll.in/article/808855/clash-of-civilisations-how-our-ancient-love-for-nature-is-losing-ground-to-modern-disregrard-for-it.

Silva, Diego. 2017. Protecting the vital: Analyzing the relationship between agricultural biosafety and the commodification of genetically modified cotton seeds in Colombia. PhD diss. IHEID.

Sindicato rural de Passo Fundo-RS. 2009. Pedido liminar. Ação coletiva. Foro Central de Porto Alegre. On file with the author.

Singh, Kshitij Kumar. 2015. *Biotechnology and intellectual property rights: Legal and social implications.* New Delhi: Springer.

Slobodian, Quinn. 2020. Are intellectual property rights neoliberal? Yes and no. *Pro-Market.* https://promarket.org/2021/04/18/intellectual-property-rights-neoliberal-hayek -history/.

Sood, Jyotika. 2013. Biopiracy case turns intense. *Down to Earth.* February 28. https:// www.downtoearth.org.in/news/biopiracy-case-turns-intense-40323.

Souza Junior, Sidney Pereira de. 2012. Justiça deve estar atenta a abuso de direito de patente. *Consultor Jurídico.* December 20. https://www.conjur.com.br/2012-dez-20 /sidney-pereira-justica-estar-atenta-abusos-direito-patente.

Souza Santos, Boaventura de. 2002. *Toward a new legal common sense: Law, globalization, and emancipation.* London: Butterworths.

Sridhar, V. 2014. US intervention in Indian agriculture: The case of the Knowledge Initiative on Agriculture. *Review of Agrarian Studies* 4(2): 93–8.

Stańczak, Dawid. 2017. State–corporate crime and the case of Bt cotton: On the production of social harm and dialectical process. *State Crime Journal* 6(2): 214–40.

Stone, Glenn Davis. 2002. Both sides now: Fallacies in the genetic-modification wars, implications for developing countries, and anthropological perspectives. *Current Anthropology* 43(4): 611–30.

Stone, Glenn Davis. 2012. Constructing facts: Bt cotton narratives in India. *Economic and Political Weekly* 47(38). September 22.

Tansey, Geoff, and Tasmin Rajotte. 2008. *The future control of food: A guide to international negotiations and rules on intellectual property, biodiversity and food security.* London: Earthscan and IDRC.

Tarragô, Piragibe dos Santos. 2015. Negotiating for Brazil. In *The making of the TRIPS Agreement: Personal insights from the Uruguay Round negotiations,* ed. Jayashree Watal and Anthony Taubman, 239–56. Geneva: World Trade Organization.

Tosi, Marcos. 2018. Monsanto pode perder patente que rende R$2,3 bilhões por ano. *Gazeta do Povo.* January 26. https://www.gazetadopovo.com.br/agronegocio /agricultura/soja/monsanto-pode-perder-patenteque-rende-r-23-bilhoes-por-ano -0kaslxljna2cc2372ahpm9ia4/.

Tubino, Najar. 2013. A lagarta que comeu o agronegócio. *Carta Maior.* April 12. https://www.cartamaior.com.br/?/Editoria/Mae-Terra/A-lagarta-que-comeu-o -agronegocio/3/27910.

UNI. 2007. Ryotu Sangham urges CM to sell BT cotton seeds at cheaper rates. *One India.* March 17. https://www.oneindia.com/amphtml/2007/02/26/ryotu-sangham -urges-cm-to-sell-bt-cotton-seeds-at-cheaper-rates-1174119458.html.

United Nations. 1992. The Convention on Biological Diversity of 5 June 1992. 1760 U.N.T.S. 69.

United Nations. 1999. *Human Development Report*. United Nations Development Programme. New York: Oxford University Press.

United Nations. 2018. United Nations Declaration on the Rights of Peasants and Other People Working in Rural Areas. A/C.3/73/L.30.

UPOV (International Union for the Protection of New Varieties of Plants). 1991. International Convention for the Protection of New Varieties of Plants of December 2, 1961, as Revised at Geneva on November 10, 1972, on October 23, 1978, and on March 19, 1991.

UPOV (International Union for the Protection of New Varieties of Plants). n.d. https://www.upov.int.

Van Brunt, Jennifer. 1985. *Ex parte Hibberd*: Another landmark decision. *Nature Biotechnology* 3: 1059–60.

Van Dycke, Lodewijk, and Geertrui Van Overwalle. 2017. Genetically modified crops and intellectual property law: Interpreting Indian patents on Bt cotton in view of the socio-political background. *Journal of Intellectual Property, Information Technology and Electronic Commerce Law* 8(151): 151–65.

Venkateshwarlu, K. 2006. Monsanto directed to reduce Bt cotton price. *The Hindu*. May 12. http://www.thehindu.com/todays-paper/tp-national/tp-andhrapradesh/mons anto-directed-to-reduce-bt-cotton-price/article3132046.ece.

Venkateswarlu, Davuluri. 2010. Seeds of child labour—Signs of hope. ILRF/ICN/Stop Child Labour.

Wattnem, Tamara. 2016. Seed laws, certification and standardization: Outlawing informal seed systems in the Global South. *Journal of Peasant Studies* 43(16): 850–67.

Wood, Ellen Meiksins. 2000. The agrarian origins of capitalism. In *Hungry for profit: The agribusiness threat to farmers, food, and the environment*, ed. Fred Magdoff, John Bellamy Foster, and Frederick H. Buttel, 23–41. New York: Monthly Review Press.

World Bank. 2006. *Intellectual property rights: Designing regimes to support plant breeding in developing countries*. Washington, DC: World Bank Agriculture and Rural Development Department.

WTO (World Trade Organization). 1994. Agreement on Trade-Related Aspects of Intellectual Property Rights. Annex IC to the Marrakech Agreement.

WTO (World Trade Organization). 1999. Minutes of the meeting of 20–21 October 1999. IP/C/M/25. https://www.wto.org/english/tratop_e/trips_e/gi1_docs_e.htm.

WTO (World Trade Organization). n.d. TRIPS: Reviews, article 27.3(b) and related issues. Background and the current situation.

COURT CASES

ADI 4234 DF. Ação Direta de Inconstitucionalidade. Supremo Tribunal Federal (2009)

Aruna Rodrigues v. Union of India. Writ Petition (Civil) no. 260/2005

Asgrow v. Winterboer. 513 U.S. 179 (1995)

Bowman v. Monsanto. 569 U.S. 278 (2013)

Cotricampo v. Monsanto. PLP no. 70010740264. Tribunal de Justiça do Rio Grande do Sul. (2005). On file with the author.

Diamond v. Chakrabarty. 447 U.S. 303 (1980)

Dimminaco AG v. Controller of Patents. (2002) IPLR 255

ESG v. NBA. Writ Petition no. 41532/2012. High Court of Karnataka at Bangalore (2012)

ESG v. NBA. Petition for Special Leave to Appeal no. 7951/2014. Supreme Court of India (2014)

ESG v. NBA. Writ Petition no. 41532/2012 (GM-RES-PIL). High Court of Karnataka at Bangalore (2013)

Executive Director v. Gridhar. Writ Petition no. 3390/2009, 2207/2008, and 2493/2010. Bombay High Court (2013)

Ex parte Hibberd. 227 U.S.P.Q. 443 (1985)

Govt of AP v. MMB. I.A. no. 05/2005 RTPE 02/2006. Monopolies and Restrictive Trade Practices Commission (2006)

J.E.M. v. Pioneer Hi-Bred. 99–1996 (2001)

MMB v. The State of A.P. Civil Appeal no. 2681 of 2006 (SC) (2006)

Monsanto v. INPI. REsp no. 1.359.965—RJ (2012/0271279-4) (2013).

Monsanto v. Nuziveedu. CS (COMM) 132/2016. Delhi High Court (2017)

Monsanto v. Nuziveedu. AIR SC 559 (2019)

Monsanto v. Sindicato rural de Passo Fundo. REsp. no. 1.243.386/RS. Superior Tribunal de Justiça (2012)

Monsanto v. Sindicato rural de Passo Fundo. no. 70049447253 (CNJ: 0251316-44.2012.8.21.7000). Tribunal de Justiça do Rio Grande do Sul (2014)

NBA v. UAS Dharwad. C.C. no. 579/2012 (PCR no. 267/2012). Court of the Chief Judicial Magistrate at Dharwad (2012)

Nuziveedu v. Monsanto. FAO (OS) (COMM) 86/2017 and 76/2017 (2018)

Monsanto v. Schmeiser. 1 S.C.R. 902, 2004 SCC 34 (2004)

Sindicato rural de Passo Fundo v. Monsanto. no. 001/1.09.0106915–2. Comarca de Porto alegre, 15a Vara Cível (2012)

Sindicato rural de Passo Fundo v. Monsanto. Recurso especial no. 1.610.728-RS (2016/0171099–9). Tribunal Superior de Justiça (2019)

UAS Dharwad v. State of Karnataka. Criminal Petition no. 10002/2013 and 10003/2013. High Court of Karnataka–Dharwad Bench (2013)

INDEX

Note: Endnote information is indicated with n and note number following the page number.

Food, Health, and the Environment

Series Editors: Robert Gottlieb, Henry R. Luce Professor of Urban and Environmental Policy, Occidental College; Nevin Cohen, Associate Professor, City University of New York (CUNY) Graduate School of Public Health

Keith Douglass Warner, *Agroecology in Action: Extending Alternative Agriculture through Social Networks*

Christopher M. Bacon, V. Ernesto Méndez, Stephen R. Gliessman, David Goodman, and Jonathan A. Fox, eds., *Confronting the Coffee Crisis: Fair Trade, Sustainable Livelihoods and Ecosystems in Mexico and Central America*

Thomas A. Lyson, G. W. Stevenson, and Rick Welsh, eds., *Food and the Mid-Level Farm: Renewing an Agriculture of the Middle*

Jennifer Clapp and Doris Fuchs, eds., *Corporate Power in Global Agrifood Governance*

Robert Gottlieb and Anupama Joshi, *Food Justice*

Jill Lindsey Harrison, *Pesticide Drift and the Pursuit of Environmental Justice*

Alison Alkon and Julian Agyeman, eds., *Cultivating Food Justice: Race, Class, and Sustainability*

Abby Kinchy, *Seeds, Science, and Struggle: The Global Politics of Transgenic Crops*

Vaclav Smil and Kazuhiko Kobayashi, *Japan's Dietary Transition and Its Impacts*

Sally K. Fairfax, Louise Nelson Dyble, Greig Tor Guthey, Lauren Gwin, Monica Moore, and Jennifer Sokolove, *California Cuisine and Just Food*

Brian K. Obach, *Organic Struggle: The Movement for Sustainable Agriculture in the U.S.*

Andrew Fisher, *Big Hunger: The Unholy Alliance between Corporate America and Anti-Hunger Groups*

Julian Agyeman, Caitlin Matthews, and Hannah Sobel, eds., *Food Trucks, Cultural Identity, and Social Justice: From Loncheras to Lobsta Love*

Sheldon Krimsky, *GMOs Decoded: A Skeptic's View of Genetically Modified Foods*

Rebecca de Souza, *Feeding the Other: Whiteness, Privilege, and Neoliberal Stigma in Food Pantries*

Bill Winders and Elizabeth Ransom, eds., *Global Meat: The Social and Environmental Consequences of the Expanding Meat Industry*

Laura-Anne Minkoff Zern, *The New American Farmer: Immigration, Race, and the Struggle for Sustainability*

Julian Agyeman and Sydney Giacalone, eds., *The Immigrant-Food Nexus: Food Systems, Immigration Policy, and Immigrant Foodways in North America*

Benjamin R. Cohen, Michael S. Kideckel, and Anna Zeide, eds., *Acquired Tastes: Stories about the Origins of Modern Food*

Karine E. Peschard, *Seed Activism: Patent Politics and Litigation in the Global South*